Environmental Footprints and Eco-design of Products and Processes

Series Editor

Subramanian Senthilkannan Muthu, Head of Sustainability - SgT Group and API, Hong Kong, Kowloon, Hong Kong

Indexed by Scopus

This series aims to broadly cover all the aspects related to environmental assessment of products, development of environmental and ecological indicators and eco-design of various products and processes. Below are the areas fall under the aims and scope of this series, but not limited to: Environmental Life Cycle Assessment; Social Life Cycle Assessment; Organizational and Product Carbon Footprints; Ecological, Energy and Water Footprints; Life cycle costing; Environmental and sustainable indicators; Environmental impact assessment methods and tools; Eco-design (sustainable design) aspects and tools; Biodegradation studies; Recycling; Solid waste management; Environmental and social audits; Green Purchasing and tools; Product environmental footprints; Environmental management standards and regulations; Eco-labels; Green Claims and green washing; Assessment of sustainability aspects.

More information about this series at https://link.springer.com/bookseries/13340

Subramanian Senthilkannan Muthu
Editor

Environmental Footprints of Recycled Products

 Springer

Editor
Subramanian Senthilkannan Muthu
SgT Group and API
Hong Kong, Kowloon, Hong Kong

ISSN 2345-7651   ISSN 2345-766X (electronic)
Environmental Footprints and Eco-design of Products and Processes
ISBN 978-981-16-8428-9   ISBN 978-981-16-8426-5 (eBook)
https://doi.org/10.1007/978-981-16-8426-5

This Springer imprint is published by the registered company Springer Nature Singapore Pte Ltd.
The registered company address is: 152 Beach Road, #21-01/04 Gateway East, Singapore 189721,
Singapore

Contents

About the Editor

Dr. Subramanian Senthilkannan Muthu currently works for SgT Group as Head of Sustainability and is based out of Hong Kong. He earned his Ph.D. from The Hong Kong Polytechnic University and is a renowned expert in the areas of Environmental Sustainability in Textiles and Clothing Supply Chain, Product Life Cycle Assessment (LCA) and Product Carbon Footprint Assessment (PCF) in various industrial sectors. He has five years of industrial experience in textile manufacturing, research and development and textile testing and over a decade's experience in life cycle assessment (LCA), carbon and ecological footprints assessment of various consumer products. He has published more than 100 research publications, written numerous book chapters and authored/edited over 100 books in the areas of Carbon Footprint, Recycling, Environmental Assessment and Environmental Sustainability.

Carbon Footprints of Recycled Plastic Packaging and Household Food Consumption by Gender in Spain

P. Osorio and M. A. Tobarra

Abstract Households' behaviour and consumption patterns are ultimately responsible for most global carbon emissions and therefore should be one main focus of sustainability policies. Detailed information provided by carbon footprints can help in that process by identifying factors and quantifying the environmental potential from changes. This chapter presents the calculated carbon footprints for Spanish single-person and average households using a multiregional input–output (MRIO) model and data from the Household Budget Survey (HBS). This allows us to analyze the differences in consumption and emissions by gender and compare it to the baseline of the average consumption unit. We find a higher carbon footprint for men (5049 vs. 4947 kg CO_2), resulting mainly from their higher level of consumption and their expending in transport, alcoholic beverages, tobacco and restaurants. This carbon footprint for single-person households is still lower than for the average consumption unit (5123 kg CO_2). When we focus on emissions from food consumption, we find however a larger carbon footprint for women as they tend to cook more at home, but double emissions from restaurants for men compared to women. Together with reducing consumption and changing patterns away from specific products and services like transport, recycling is a crucial part of the environmental strategies. In this chapter, we use data from HBS food expenditure and life cycle emissions from PlasticsEurope (The circular economy for plastics—a European overview, 2020), emissions shares from plastic packaging by Poore and Nemecek (Science 360:987–992, 2018) and total and percentages of recycled packaging from Ecoembes (Las cifras del reciclaje, 2019). We compare the carbon footprint from recycled versus new plastic and calculate the potential reductions in emissions from increasing the share of recycling for households' packaging.

Keywords Carbon footprint · Recycled products · Multiregional input–output model · Sustainability and gender · Plastic packaging

P. Osorio · M. A. Tobarra (✉)
Department of Economics and Finance, Facultad de Ciencias Económicas y Empresariales, Universidad de Castilla-La Mancha, Plaza de la Universidad, 1, 02071 Albacete, Spain
e-mail: MariaAngeles.Tobarra@uclm.es

1 Introduction

Climate change already affects all regions in the world and the time to fix it is running out. Under most emissions scenarios considered by IPCC (2021), we will not achieve the goal of keeping the rise in global temperatures under 1.5 °C. This problem will require strong, quick and sustained reductions in greenhouse gas emissions, so all the economic agents must focus on that. This global climatic emergency has raised concerns about consumption sustainability in recent years. As households are responsible for almost 75% of global carbon emissions (Druckman and Jackson 2016), it is crucial to identify potential drivers to advance towards a more sustainable economy.

Consumption patterns are the focus of recent studies both to identify their environmental impact and the policy options in order to induce changes towards reducing emissions and use of natural resources. That behaviour from consumers has been related to a different individual, household and regional determinants. In that line of research, as observed in multiple studies, gender is a key difference in environmental attitude and behaviour.

Women have been identified in several studies as being more environmentally aware and more prone to take into account the impact of their economic decisions on others and their community. Their consumption level and structure are also determined by their income, which tends to be lower than for their male counterparts and other individual characteristics such as their age, educational attainment and work status. Differences also arise from their activities, as for example, women tend to cook more rather than eat out and make more intense use of public versus private transport.

These differences can be significant when designing policies and incentives for consumers and justify delving deeper into their extent and causes. In this regard, descriptive analysis from data that includes expenditure and environmental effects by individual characteristics can become fairly useful.

Considering the above-mentioned factors, the present research analyzes carbon footprint by gender in Spanish single-person households, compared to the average household. Carbon footprints have been calculated by using a multi-regional input–output (MRIO) model that follows the seminal Wassily Leontief's input–output methodology (Leontief and Ford 1972) as well as a growing literature on the subject (Wiedmann and Lenzen 2018; Ivanova et al. 2017; Tobarra et al. 2018). This methodology allows us to combine detailed data with information on inter-industrial links between countries, disaggregating global production chains.

We start from microdata from the Spanish Household Budget Survey (INE 2015) that we process using R and SPSS software. To make these data compatible with the System of National Accounts (SNA), we follow Cazcarro et al. (2020). Data homogeneity avoids potential biases and is achieved by reorganizing and transforming the original data to take into account differences between products and industries, classifications for HBS and SNA and basic versus purchaser prices.

Changes in consumption patterns and environmental behaviour are one of the strategies put forward to address climate change challenges. In this sense, recycling has become a crucial part of national sustainability strategies. To shed some light on this issue, we combine data on food expenditure from the HBS, information on life cycle emissions from PlasticsEurope (2021), percentages that packaging represents for different food products (Poore and Nemecek 2018) and shares of recycled packaging by households for Spain (Ecoembes 2021).

The calculated footprints are used to analyze the differences in consumption and emissions patterns by gender. Our results showed a relevant difference between men and women, with the former having a higher consumption level (3%) that translates into a higher carbon footprint (2%). Sector analysis showed men double women's consumption of alcoholic beverages and restaurants. Women, on the other hand, spend 12% more on food. Then, the consumption level translates into a higher carbon footprint. An average household footprint is 5123 kg CO_2, whereas it is 4947 kg CO_2 for women's households and 5049 kg CO_2 for men.

As for emissions from food packaging, even though women living alone spend more on food than men, we find a more significant role for packaging in male emissions. This is dependent on food consumption patterns, as men buy more wine, coffee and other beverages, which are linked to greater emissions from packaging. That also highlights the need for increasing recycling in Spain. Compared to countries like Germany, where all plastic packaging is either recycled or used for energy recovery, 34% of plastic packaging in Spain ends up in a landfill (PlasticsEurope 2021). We also calculate, using different estimates for plastic production and recycling as those by CIEL (2019), how much increasing the share of recycled packaging may lower households' emissions.

In the following sections, we develop all the required steps. We will start by reviewing the recent literature on the relevance of carbon emissions, particularly from households' consumption and the importance of individual characteristics and expenditure patterns. We also comment on recent articles and data sources for emissions from plastics and recycled packaging products. We will then detail the methodology used in this chapter, including the multiregional input–output (MRIO) environmentally-extended model, HBS data and the transformations required to make those data compatible and calculate the carbon footprints and data sources and calculations to obtain the emissions incorporated in all plastic packaging and particularly, recycled plastic packaging from households. Chapter 4 shows our main results (carbon emissions from household consumption, with particular detail for food products, emissions from food packaging and emissions from recovered plastic packaging, specifically recycled plastic), while Chap. 5 concludes.

2 Literature Review

Since the 1970s, the world's population has doubled and the global GDP has quadrupled, while the use of natural resources has tripled (PIR 2019) and carbon dioxide levels have grown by 26% (WMO 2019). In Spain, there has been a cumulative increase in the average temperature of 1.3 degrees since the 1960s, according to AEMET (2019). It is apparent that climate change is already, directly and indirectly, affecting all economic sectors and all ecological systems in this area (Government of Spain 2019).

Despite the economic downturn caused by the pandemic and a temporary reduction of emissions, the concentration of greenhouse gases (GHG) continues to grow (WMO 2020). In addition, the latest IPCC report shows the existing global threats arising from climate change: increasingly frequent extreme weather events (heatwaves, heavy rainfall, droughts), melting glaciers, sea-level rise, etc.

In this regard, the reduction of greenhouse gases (including carbon dioxide) is considered a top priority, according to the 2030 Agenda. The latest proposal from the European Commission aims to reduce greenhouse gas emissions by at least 55% before 2030. In addition, its twelfth objective promotes the efficient use of resources by firms and consumers and suggests that large companies should prepare reports on sustainability. With that in mind, ecological footprints (such as carbon footprint) are valuable tools for assessing sustainability.

The concept of "carbon footprint" refers to the amount of carbon dioxide (CO_2) released into the atmosphere, derived from activities of production or from the consumption of goods and services (Pandey et al. 2011; Hoekstra and Wiedmann 2014). Calculations and analyzes on the carbon footprint can be approached from different perspectives, choosing between different economic agents. For example, it could be used to study the emissions of a country, such as Spain (Cansino et al. 2012), an institution (FEMP 2014; Gómez et al. 2016), a congress (GEAR 2020), or a specific economic agent, such as households.

The carbon footprint of households has previously been analyzed by López et al. (2016) and Arce et al. (2017) for Spain, by Gao et al. (2020) for China, by Druckman and Jackson (2010) for the United Kingdom, among many others.

Delving into the characteristics of the household as determinants, environmental studies by gender are very limited. Two of them are the main reference within this field: Toro et al. (2019) and Druckman et al. (2012). This shows that further research on this topic is required.

It is relevant to know who the most polluting agents are, but it is also important to know which sectors contribute the most to these emissions. Different studies state that most of the carbon footprint of a household corresponds to the transport sector (cars, motorbikes...), the housing sector and the food sector (Druckman and Jackson 2016; Räty and Carlsson-Kanyama 2010).

As for the food sector, it is a crucial component because it represents a high percentage of household expenditure and its composition has a relevant impact on emissions. Furthermore, it is a relatively sensitive part of expenditure in terms of changes in behaviour and trends.

In Spain, the carbon footprint of food (taking into account all stages) has grown fourfold in total terms, between 1960 and 2010 (Aguilera et al. 2020). Globally, a recent study shows that emissions from the food system account for 34% of total GHG emissions in 2015, the highest contribution coming from agriculture and land use activities (71%) and the remaining coming from retail, transportation, consumption, fuel production, waste management, industrial processes and packaging (Crippa et al. 2021).

Other studies reach different values: 26% of greenhouse gas emissions corresponds to food sector, according to Poore and Nemecek (2018), whereas IPCC estimated that emissions from food systems may account for 21–37% of total emissions (combining previous literature) (Mbow et al. 2019).

Focusing on households, not all dietary patterns have the same impact. It is a fact that, compared to vegetables, animal products have a greater climate and land use impacts (Nijdam et al. 2012). In Spain, 80% of emissions from food production correspond to food products of animal origin, especially pork, beef, milk and fish (Aguilera et al. 2020). Taking all processes in the food system into account, Poore and Nemecek (2018) estimate that 58% of all emissions from the food system are related to animal products.

Transitioning toward diets that exclude animal products is crucial to reducing GHG emissions. In particular, greenhouse gases could be reduced by a range within 29–70%, compared with a reference scenario calculated by Springmann et al. (2016). Returning to Poore and Nemecek again, they evaluate the potential reduction to be 49%.

Concerning diet changes, attitudes towards meat consumption and the preference for vegetable products vary by gender, according to a report that reviewed previous literature on this topic (Modlinska et al. 2020). Some papers by Carlsson-Kanyama highlight the differences in eating patterns between men and women.

First, a study by Carlsson-Kanyama et al. (2003) calculated the energy requirements "from farm to table" of food products and studied the differences in the consumption of men and women in Sweden. The results showed that energy intakes were up to 21% higher in men's food consumption than women's and men's higher meat consumption partly explains this difference. Second, in a later study (Räty and Carlsson-Kanjama 2010), the differences in consumption patterns were re-emphasized after analyzing four European countries. The results for dietary patterns show that women require more energy in their consumption of fruits and vegetables and men in meat products. Also, men stand out for their greater recourse to eating out, so a relevant part of the food energy consumption comes from restaurant and hotel services.

Some research developed by regional governments in Spain shows that men tend to eat out more and consume more significant amounts of alcohol, tobacco and meat or dairy products, while women consume more fruit, vegetables and cereals (Red2Red Consultores 2014; Dirección General de la Mujer 2014). Likewise, women adopt a vegetarian or vegan diet in Spain (Lantern 2019) more often than men. In addition, the OECD (2008) shows that women are more likely to recycle and buy organic food.

It is important to understand eating patterns and change polluting habits, but this is not the only and perfect solution. Furthermore, the food system is more than the production and consumption phases. According to FAO (2018, p. 1), "the food systems encompass the entire range of actors and their interlinked value-adding activities involved in the production, aggregation, processing, distribution, consumption and disposal of food products that originate from agriculture, forestry or fisheries and parts of the broader economic, societal and natural environments in which they are embedded".

As part of that process, we should consider packaging and the related emissions. Food packaging provides protection and prevents food contamination, thus extending the shelf life of the product (Singh et al. 2017). The choice of material (glass, paper, metal, plastic) will determine the design, quality, safety and many other properties of the food product. An essential finding of the study of Verghese et al. (2019) is that packaging can have a significant impact on reducing food waste in the food supply chain and achieving sustainability.

Plastic is the most commonly used material for packaging, as it has the required properties in terms of hygiene and the advantage of relatively low cost and weight, allowing a reduction in food waste and the cost of transportation. Similarly, packaging is one of the most used applications of plastic (Geyer et al. 2017). Globally, it represents 26% of the total volume of plastics worldwide (WEF 2016). In the European Union, it amounts to nearly 40% of total plastic production and 61% of plastic waste (PlasticsEurope 2020). Besides, almost half of the plastic used for packaging is used for food products or beverages (ING 2019a).

Regardless of the positive points of plastics as packaging material, the disadvantages of their use cannot be forgotten. Plastics are mainly derived from crude oil (Geyer et al. 2017), so large quantities of CO_2 are released during their incineration (together with other pollutants such as nitrous oxide or mercury), according to an EU report (2020). Greenhouse gas emissions from plastics could account for 15% of the annual global carbon budget by 2050 (WEF 2016).

Furthermore, the problem of plastics goes beyond air pollution: this report points that sometimes plastics end up in the sea, resulting in around 85% of the litter found on European beaches being plastic waste. Nowadays, plastics and microplastics can be found everywhere: even in the depths of the Mariana Trench. It takes hundreds of years for plastics to disappear, increasingly in the oceans, where if nothing changes, there will probably be more plastic than fish by 2050 according to WEF (2016).

An international survey found, from a sample of more than 13,000 individuals, that plastic waste was listed at the top of their environmental concerns for 34% of respondents, compared to 33% for climate change or 14% of air pollution or 11% for depletion of natural resources (ING 2019b). For the specific case of Spain, 37% saw plastic waste as the most pressing environmental issue, only surpassed by the 42% that identified climate change as the biggest problem. This shows a large awareness in Spain for these two issues and justifies our consideration of both emissions and plastic waste in this chapter. Not only that, but 80% of Spanish participants in the survey were willing to give priority to the environment above economic growth and 75% identified over-consumption as a relevant contributor to these problems.

The European Commission, which is aware of the problem and increased public awareness, has reviewed the targets for packaging and set more demanding rules. To guarantee that all packaging in the EU market is reusable or recyclable by 2030, the updating of the Packaging and Packaging Waste Directive (PPWD) has introduced tighter criteria for calculating recycling rates, which means a significant reduction in reported recycling rates (from 42 to 29%). Besides, recycling targets have changed in order to reach 50% by 2025 and 55% by 2030, which means that ambitious progress is still needed to accomplish this new goal.

Recycling reduces plastic emissions (Dormer et al. 2013) and it is an easy way to start being sustainable as individuals. However, even though there have been multiple information campaigns, there are still many people who decide not to recycle (most often, because they lack space in their kitchens or waste containers are far or insufficient, IPSOS 2019). As for Spain, a survey shows that 80% of the population usually recycles plastic, glass and paper waste. If we take a look at gender differences, women are more concerned with separating and not generating unnecessary waste (83%) than men (78%) (IPSOS 2019). Focusing on packaging (plastic, metal, paper, etc.), the recycling rate in Spanish households is almost 79%. If we examine the rate of household package recycling by type of material, the recycling rate for plastic reached 75.8%, compared to 85.4% for metal and 80% for paper (Ecoembes 2021).

A positive aspect can be that this increase in recycling rates shows a greater interest by the population for the sustainability of consumption, but there is a long road ahead. There are not enough studies that analyze the impact of gender on this issue (as far as we know). Considering the global problem of plastics and its impact on climate change, it is relevant to analyze behaviour patterns (such as the effect of gender or age).

3 Methodology

3.1 Data Collection

3.1.1 Carbon Footprints Calculation

The microdata used has been obtained from the Household Budget Survey (HBS), provided by the National Institute of Statistics (INE 2015). The HBS data allow us to know the consumption expenditure of households in Spain through a sample of 24,000 households.

Data have been collected following two main criteria. On the one hand, we aim to use the most up-to-date data available. On the other hand, we need to take into account that we have to combine data from different sources for the estimation process and therefore, decisions about the time period and product classification of the HBS data had this in mind. The consumption expenditures considered in this study include the monetary flow of final consumption of goods and services and the value of non-monetary consumption by households.

Once the three HBS microdata files as provided by INE were downloaded, the information was loaded and merged using the IBM SPSS Statistics software. Afterwards, the relevant variables were selected and, for the study of men and women, dummy variables were created. Single-person households have been selected because this allows our results to be as homogeneous and comparable as possible, avoiding problems related to the composition of the household or related to the selection of the main breadwinner. In similar studies (Toro et al. 2019; Räty and Carlsson-Kanyama 2010), the same type of household was used to identify consumption patterns by gender.

Household consumption is grouped into the 3-digits COICOP categories (Classification of Individual Consumption by Purpose) and the following vectors are obtained: (1) expenditure vector of Spanish households; (2) expenditure vector of male single-person households; (3) expenditure vector of female single-person households.

Next, the data have been transformed to make them compatible with the input–output (I-O) approach, following the methodology by Cazcarro et al. (2020), using data from the NA for population (INE 2020) and the supply table at basic prices for 2015 (INE 2018). In addition, the expenditure vectors were transformed from euros to US dollars using the exchange rate provided by WIOD (Timmer et al. 2015, 2016), as this database provides data in that currency.

Finally, carbon footprints are computed based on the multi-regional input–output table, which is provided by the 2016 WIOD database (Timmer et al. 2015, 2016) and the emissions data (in CO_2 kilotons) from the EC Joint Research Centre (Corsatea et al. 2019). The I-O table and the emission vector are limited: they relate to the last year available, 2014, while our HBS information dates to 2015. However, this limitation of the available data is not expected to introduce any significant bias in our results, under the assumptions of constant technology and fixed commercial structure.

The footprints have been calculated using the MATLAB software (MathWorks Inc., 2016), by appropriately adapting the code developed by Monsalve (2017).

3.1.2 Food, Packaging, Recycled and Life Cycle Emissions

The HBS data have also been used to obtain detailed information for food expenditure in Spanish households by gender and age. From the food expenditure categories, the physical quantities consumed of each category (expressed in kilograms) can be calculated. This could be done by using the prices of each product in 2015, which were downloaded from the Ministry of Agriculture, Fisheries and Food (2021), so physical quantities were calculated by dividing the expenditure by the corresponding price by kilogram or litre. Emissions from food packaging have then been calculated for these products categories, following the estimates of Poore and Nemeck (2018). We have also used information on life cycle emissions from PlasticsEurope (2021) and shares of recycled packaging by households for Spain (Ecoembes 2019).

In order to provide additional information on the role of recycling on the GHG emissions issue for household consumption in Spain, we combine information from different sources to provide some light on the amount of packaging used by consumers and, more precisely, the impact of plastic packaging waste.

We will firstly use data from Ecoembes (2021) for collected household packaging waste in Spain and from our calculation of the number of equivalent consumption units in Spanish households, we can obtain the amount (in physical units, kg) of waste packaging made of different materials per equivalent consumption unit. Even though we can also offer those amounts in per capita terms, we find it more relevant and consistent with our carbon footprint calculations, to use this measure.

From those amounts of packaging waste, we will focus on plastic, as it is one of the major air and water polluters and it is rapidly increasing, compared to other waste products. We use data from Zheng and Suh (2019) and the Center for International Environmental Law (CIEL 2019), on GHG emissions for production and end-of-life stages for plastic. While Zheng and Suh (2019) aim to provide averages for world production of plastic, CIEL rather focuses on North American and European data. As the figures of both sources differ (in terms of energy mix, share of plastic types, etc.), we will offer the results from calculations using both sets of data, so we are aware of this source of uncertainty.

Finally, we will use data from the Ministry of Ecological Transition and Demographic Challenge, which collects information from the Integrated Systems of Waste Treatment. These are also the data that PlasticsEurope considers in their statistics for Spain. We will also show the difference with other sources, like Cicloplast (2019). The information provided allow us to calculate the share of collected waste that is either recycled or destined for energy recovery or disposal in landfills. Combined with the previously described data by Zheng and Suh (2019) and CIEL (2019), we will calculate the GHG linked to the end-of-life treatment of household plastic packaging waste, both in the current situation and considering the objective by the European Parliament of recycling 55% of all this material by 2030.

3.2 Data Homogeneity

To make these data compatible with the System of National Accounts (SNA), we follow Cazcarro et al. (2020). Data homogeneity avoids potential biases and is achieved by reorganizing and transforming the original data so that the different classifications and the differences between basic versus purchaser prices are taken into account. Four steps were implemented:

1. Aligning survey data to NA principles (adjusting the population value, uprating/downrating categories and adding the value of prostitution and narcotics).
2. Converting the data classified by COICOP (Classification of Individual Consumption by Purpose) to CPA (Classification of Products by Activity) using a bridge matrix.
3. Adjusting the valuation from purchaser's prices to basic prices. This involves detracting net taxes (taxes less subsidies) and then separating and reallocating trade and transport margins to retail and transportation sectors.

4. Adapting the data from a product to an industry-based classification, as the input–output table we employ, WIOD, shows the links between different countries and sectors on an industry-by-industry basis.

We provide a more detailed description of these four steps in the remaining of this section.

3.2.1 Alignment of Survey Data to NA Principles

Several adjustments are made in order to align the expenditure vector obtained from the HBS microdata with the National Accounts (NA) principles. It should be noted that the downloaded data could be substantially different from the NA data (Zwijnenburg 2016). Some of the special features are detailed below.

First, the population value of the HBS (which does not coincide with the population census used by NA) must be adjusted by using a correction factor. This is due to the fact that some types of dwellings fall outside the HBS but are included in the National Accounts (institutional dwellings such as prisons or hospitals). In our specific case, we compare the population data extracted from the HBS microdata for 2015 with the information available in the INE population census for the same year (INE 2020).

Second, some of the HBS COICOP categories must be re-adjusted, to fully match the COICOP-NA categories. This is the case as there are some expenses that households do not report, such as the so-called FISIM (Financial Intermediation Services Indirectly Measured), which are accounted for in the CNE but not in the HBS (households are usually not aware of this margin with which they are implicitly charged by the banks). There are also other expenditures not included in the HBS but measured in the CNE, such as social transfers in kind (e.g. education or health). Besides, the categories corresponding to narcotics and prostitution are underestimated or non-existent in the HBS because households do not report them as much as they consume (something that also happens to a lesser extent with tobacco and alcoholic beverages). By contrast, usual household expenditures such as food and miscellaneous items are often overestimated. However, the National Accounts do take all this into account when assessing the production of each sector.

As a solution, the HBS values in each category must be revalued (uprate or downrate, as needed), according to the existing gap between the survey and the CNE. Specifically, the resulting total HBS data for each product must be compared to the NA's Household Final Consumption Expenditure in COICOP categories (INE 2013) so a set of coefficients that bridge the gap between both sources is constructed. Once these values are obtained, each expenditure vector (3-digit COICOP) for individuals or average type of household should be downrated or uprated by multiplying them by these coefficients.

The limitations of this adjustment must be considered. By applying a national average to each vector, the reality is being simplified. That is, all households, despite their characteristics, are adjusted in the same way since there are no data to account for the methodological differences in the way the information is collected and provided by the two sources to allow for more appropriate correction coefficients. In conclusion, this adjustment does not solve the problem completely, although it helps to reduce the issues of using data reported by households and their methodological discrepancies with the NA.

3.2.2 Conversion of Data Classified by COICOP to CPA

After the alignment, the data classified by the type of product according to the purpose (COICOP) must be converted to CPA in the 2008 version. The Classification of Products by Activity (CPA) is the European version of the Central Product Classification (CPC) prepared and recommended by the United Nations Organization (INE 2009).

Bridge matrices can be constructed to this aim. In this work, we start from the bridge matrix provided by Cazcarro et al. (2020), which presents the CPA products in rows and the COICOP products in columns. This matrix was, however, developed for the year 2010, while the HBS data used in this study are from 2015. In order to correct for changes between those two dates, we apply the RAS[1] method that, starting from a matrix and data on the sum of columns and rows, provides a new matrix that replicates as closely as possible the original structure but adapting it to the new totals (Miller and Blair 2009). To this aim, we use the code for MATLAB by Temurshoev (2013).

3.2.3 Adjustment of the Valuation from Purchaser's Prices to Basic Prices

The data from the previous step are valued at purchaser's prices (PP), so this data was revalued to reflect basic prices (BP). According to the United Nations-System of National Accounts (United Nations 2009), purchaser's prices includes taxes minus subsidies on products (TLS) (excluding deductible taxes such as VAT), the trade margins, transportation charges (TTM) and deductions for any discounts.

This implies that, for example, HBS food expenditure includes the money paid for taxes (which is not produced in that sector and therefore should not be included) as well as the amount paid for intermediation and transportation of that food (which should go, respectively, to the trade and transportation sectors and not to the food sector).

The final consumption vector from the previous step is revalued from purchaser's prices to basic prices when taxes are subtracted and transport and trade margins are reallocated to the sector to which they belong (1):

---

[1] RAS ("raking and scaling") is an algorithm that adjusts matrices by repetitively multiplying the elements in rows and columns in a base matrix by correction factors provided by exogenous information, normally from a different time period that the original matrix, in order to reconcile the information from both sources and obtain an updated matrix.

$$PP = BP + TTM + TLS \tag{1}$$

This follows the definitions of the United Nations-System of National Accounts (United Nations 2009). Basic prices exclude taxes (TLS), trade margins and transport costs (TTM).

Unlike other European countries, these tax and margin data for the Spanish economy are not publicly available, so we draw on the estimation by Cazcarro et al. (2020).

3.2.4 Adapt the Data from Product Classification to the Industry Standard

In this last step, the data must be adapted to the specific input–output format by means of the so-called "Model D". If the I-O table to be used is in the industry-by-industry format, as in this case, the consumer profile of the vector (obtained in the previous step) must be adjusted from a product classification (CPA) to an industry classification (ISIC). In other words, we need to adapt the data from a use (non-symmetric) product by industry table to a (symmetric) industry-by-industry I-O format [ISIC Rev.3 (Timmer et al. 2015, 2016)].

According to the recent handbook of national accounting on the topic of input–output tables by the United Nations (Mahajan et al. 2018), there are different available options to allocate product data to the corresponding industry. In doing so, all of them aim to account for secondary production (subsidiary products, by-products and joint products). These options depend on assuming a fixed product or industry technology or a fixed product or industry sales structure. Among those, we follow Cazcarro et al. (2020) in using Model D to perform the adaptation towards the industry-by-industry format. Applying this Model D involves the assumption of fixed products sales structure,[2] so it is necessary to adjust the rows in such a way that we obtain product-adapted industries in the rows by transferring inputs within the industry columns. Finally, the product classification of the rows is then transformed into the industry classification (industry-adjusted products) of the columns.

For this purpose, the transformation matrix is calculated using the supply table at basic prices, in which the products are shown by rows and the industries by columns. Basically, this transformation matrix is the value of each element of a row in the supply table divided by the total sum of that same row. In this way, we calculate how much output corresponds to the production of an industry (of the total output of a product category).

That matrix is then transposed and multiplied by the expenditure vector for each household, as defined above, so we end up with a vector of final demand in terms of 64 industries (ISIC Rev. 4). The last step involves aggregating data for some of those

---

[2] Each product has its own specific sales structure, irrespective of the industry where it is produced.

industries, as the international I-O database we use for calculating the carbon footprints (WIOD, as described in the following section) provides data for 56 industries (ISIC Rev. 4).

3.3 Carbon Footprints

Once the appropriate transformations have been implemented on the expenditure data, so it is compatible with the I-O database, carbon footprints are calculated. Carbon footprints are computed based on the multi-regional input–output model. To begin with, input–output tables are a technique of macroeconomic analysis based on National Accounts that show the interdependencies between different economic sectors and different economic areas. Our model is called *Environmentally-Extended Multi-regional Input–Output* (EEMRIO) as it includes information on environmental data with the same level of disaggregation so not only can we obtain all intermediate inputs, directly and indirectly, involved in producing each particular amount of output, but we can also calculate the related factor requirements. This can be used for labour, for example, but also for natural resources, such as water, or for waste products, like GHG emissions. Therefore, this allows us to rewrite the original equation and use it for environmental calculations.

Input–output models distinguish two categories of traded products in an economy (be it at regional, national or global levels): trade in intermediate goods acquired by industries and trade in final goods that supply final demand. The most basic model is represented by the following expression (2) (Martínez and Roca 2013):

$$Y = X - (A \cdot X) = (I - A) \cdot X \tag{2}$$

A is the matrix of technical coefficients (quantities of each product used as an intermediate input per unit of output for each supplying industry and each producing industry); X is the vector of total output; $(A \cdot X)$ are the inputs needed to produce X; I is the identity matrix and Y is a vector of final demand. We can rearrange it to obtain (3):

$$X = (I - A)^{-1} \cdot Y \tag{3}$$

This fundamental I-O equation shows that the amount of goods produced depends on the (constant) technology expressed in the technical coefficients and the final demand. In its simplest form, it is used for calculations in a single region/country. Still, it is extensible to a multi-regional model that includes productive and demand links between many sectors and regions.

As explained above, this equation allows us to calculate the impact of the demand and production of goods (in terms of other variables). For example, the environmental impact or social impact from a region's final demand can be analyzed. The generic expression in Eq. (3) is adapted to the specific model by integrating a satellite account—in this case, CO_2 emissions—and this combination results in the so-called environmentally-extended input–output (EEIO), which extends the analytical capacity of the initial model (Miller and Blair 2009). The resulting expression is as follows (4):

$$E = \hat{e} \cdot L \cdot \hat{y} = ME \cdot \hat{y} \qquad (4)$$

where $\hat{e}$ is the diagonalized vector of direct CO_2 emissions per unit of output in each sector of each country or region; L is the Leontief inverse matrix $(I - A)^{-1}$ calculated from the identity matrix I and the matrix of technical coefficients A; $\hat{y}$ is the diagonalized matrix of final demand for each product/sector and country. Multiplying the vector of emissions by the Leontief inverse matrix provides the matrix of emission multipliers by country and industry (ME). This matrix concentrates all the relevant information about which products/industries and countries generate the emissions incorporated per euro in each type of product demanded by each country.

The technical coefficients A are defined as direct input ratios, which show the amount of input from sector i needed to produce one unit of output from j (inputs per unit of output). Going deeper, this matrix of coefficients decomposes into the matrix of domestic production for region r (A_{rr}) and into the matrix of imports from each region r to each region s (A_{rs}). This matrix A allows us to calculate the Leontief inverse matrix L, which shows the required production (direct and indirect input) of sector i from each region r to produce one unit of output of sector j in region s and therefore supply the final demand of all products and regions.

To illustrate this, we express Eq. (4) in matrix form for a simplified economy, assuming two regions (r, s) and two productive sectors (i, j):

$$
\begin{pmatrix}
E_{ii}^{rr} & E_{ij}^{rr} & E_{ii}^{rs} & E_{ij}^{rs} \\
E_{ji}^{rr} & E_{jj}^{rr} & E_{ji}^{rs} & E_{jj}^{rs} \\
E_{ii}^{sr} & E_{ij}^{sr} & E_{ii}^{ss} & E_{ij}^{ss} \\
E_{ji}^{sr} & E_{jj}^{sr} & E_{ji}^{ss} & E_{jj}^{ss}
\end{pmatrix}
=
\begin{pmatrix}
e_i^r & 0 & 0 & 0 \\
0 & e_j^r & 0 & 0 \\
0 & 0 & e_i^s & 0 \\
0 & 0 & 0 & e_j^s
\end{pmatrix}
\begin{pmatrix}
L_{ii}^{rr} & L_{ij}^{rr} & L_{ii}^{rs} & L_{ij}^{rs} \\
L_{ji}^{rr} & L_{jj}^{rr} & L_{ji}^{rs} & L_{jj}^{rs} \\
L_{ii}^{sr} & L_{ij}^{sr} & L_{ii}^{ss} & L_{ij}^{ss} \\
L_{ji}^{sr} & L_{jj}^{sr} & L_{ji}^{ss} & L_{jj}^{ss}
\end{pmatrix}
\cdot
\begin{pmatrix}
y_i^{rr} & 0 & y_i^{rs} & 0 \\
0 & y_j^{rr} & 0 & y_j^{rs} \\
y_i^{sr} & 0 & y_i^{ss} & 0 \\
0 & y_j^{sr} & 0 & y_j^{ss}
\end{pmatrix}
$$

$$(5)$$

Read by columns, the resulting expression show the carbon footprint (López et al. 2017), as it indicates all emissions incorporated in all inputs (for all industries and countries of origin of those inputs) required to produce the final demand for that product and region.

The multiplier ME shows the requirements of that factor to obtain one unit of output and it is obtained premultiplying the Leontief inverse by the emissions factor. Finally, the ME matrix is multiplied by the final demand diagonalized by sectors (in this case, it corresponds to household consumption). The result is matrix E (CO_2 emissions measured in kilotons), where superscripts show the relationships between countries and subscripts show the links between sectors. We used MATLAB to cope with the complexity of matrix calculations (56 sectors and 44 regions).

Matrix E can be interpreted from two perspectives, one by columns and the other by rows. By columns, the result obtained is the consumers' responsibility (or carbon footprint); the emissions resulting from the final demand for goods and services. By rows, the result is the producers' responsibility (in a region) in the production process, resulting from their demand for inputs to generate final goods. In summary, by analyzing the results by columns and rows, we obtain a view in terms of final goods demanded or in terms of sectors/regions in which emissions are originally generated.

4 Main Results

4.1 Carbon Footprint and Emissions Pattern

In this section, we present the HBS data on expenditure by product category, with special attention to consumption by the average Spanish household compared to single-person (male and female) households. This is an essential basis for the following analysis on carbon footprint, as both the level and composition of their demand greatly affect the results.

First of all, there is more single-person female than male households (2,502,565 vs. 2,194,886) and they represent, respectively, 14% and 12% of the total number of Spanish households. In addition, the total expenditure of female single-person house-holds is higher than that of male households (52,729,007,598€ vs. 47,551,591,663€) in 2015.

Aggregate data for the whole economy show that male single-person house-holds spend less on food and non-alcoholic beverages but more on alcoholic bever-ages and restaurants. Specifically, total male consumption is 4,749,909,401€ versus 6,313,481,097€ for women in the food category. In restaurants, respectively, the expenditure is 7,954,479,856€ and 4,164,557,316€ in 2015.

To concentrate on differences among households, rather than the total levels for all households, we will proceed to analyze consumption and carbon footprint in terms of the average household. Specifically, spending will be contrasted between the average Spanish household (expressed in equivalent consumption units[3]), the average male single-person household and the average female single-person household.

The expenditure per equivalent consumption unit divides the expenditure between the people living together but takes into account the existence of economies of scale. We speak of economies of scale in households when some goods or services are

---

[3] Equivalent consumption units allow us to compare consumption per capita taking into account the fact that part of the consumption for a household is independent from the number of its members. They are calculated using the OECD modified scale that assigns the following weights: 1 for the first adult, 0.5 per any other adult, and 0.3 for each under-14 individual. Once we obtain the expenditure per equivalent consumption unit in each household (dividing the total expenditure by the number of units), we will allocate it equally to each member.

enjoyed by several members without the need to spend a larger amount (Ortiz and Marco 2006).

Taking the other categories apart from the food sector (transport, housing, education, health…) into account and following the steps detailed in the methodology, we can present here results for global carbon footprints and their disaggregation by type of product. All Spanish households generate 157,900 kt of CO_2 and emissions for the average household are 5123 kg of CO_2 per year.

If we look at single-person households, the generation of emissions is 12,380 kt of CO_2 for all-female households and 11,083 kt of CO_2 for male households, which represent almost 8% and 7% of total emissions, respectively. On average, a female single-person household generates 4947 kg of CO_2 compared to 5049 for a man, which is still slightly lower than the average household. These data show a difference in carbon footprint by gender of 2%.

Delving into the emissions by food product, we calculated the total amount generated (measured in kg of CO_2 equivalent per kilogram of product) by using the proportions estimated by Poore and Nemecek (2018). Emissions of food products include the greenhouse gases emissions of each stage in the supply chain: land use, animal feed, farm, retail, processing transport and packaging. Within these emissions, we will focus on packaging emissions, which include the production of the packaging material, transportation of the material and end-of-life disposal (Fig. 1).

The chart above shows food emissions of an average household in Spain (2015) in the products selected. The most polluting food products come from animals. First, the supply chain for meat (beef, lamb, pork and poultry meat) generates more than 600 kg of CO_2e per kg of product. Second, milk generates 428 kg of CO_2e, followed by cheese, fish and shellfish. Total emissions add up to 2331.5 kg of CO_2e per kg of product in an average household.

Taking a look at emissions from single-person households, those are higher for almost all products for a female household than for a male household. However, a man stands out in alcoholic beverages versus a woman and compared to the average household (Table 1).

First, an average woman generates more emissions in the supply chain (all embodied emissions in the production of a particular final product that she demands) of sugar, vegetables, pulses and legumes, oils, coffee, meat and other animal products (fish, shellfish, cheese, eggs, etc.) than a man. However, an average man generates 83% more emissions from the supply chain of beer (34 kg of CO_2e) and 46% more from wine (close 16 kg of CO_2e more). In summary, for the products selected in this study, emissions are lower in an average male household than in an average female one (16% less).

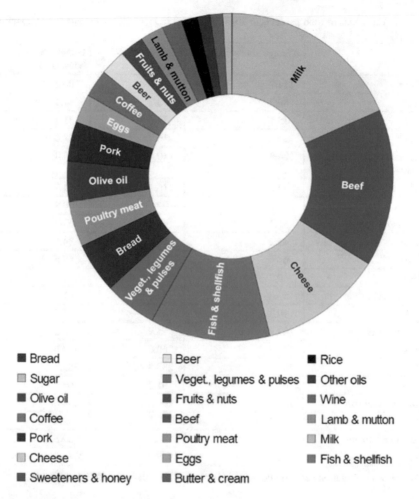

■ Bread	□ Beer	■ Rice
■ Sugar	■ Veget., legumes & pulses	■ Other oils
■ Olive oil	■ Fruits & nuts	■ Wine
■ Coffee	■ Beef	▨ Lamb & mutton
■ Pork	■ Poultry meat	▨ Milk
▨ Cheese	■ Eggs	■ Fish & shellfish
■ Sweeteners & honey	■ Butter & cream	

Fig. 1 Distribution of emissions of food products generated by average Spanish household in 2015 (measured in kg of CO_2 equivalent per kilogram of product) *Source* Own elaboration based on Poore and Nemecek data (2018) and INE data (2015)

4.2 Consumption Patterns: Food, Restaurants and Age Group

By analyzing the expenditure profile of the average household, we find that a woman spends a higher percentage of her expenditure on food and non-alcoholic beverages (13%) than a man (11%). More precisely, an average woman spends 2715€ while a man spends 2369€. This represents a gender gap of 13%.

Meanwhile, a man spends 3624€ in restaurants per year, which is 17% of his total expenditure (and it exceeds the expenditure on food) and women spend only

Table 1 Emissions of food products in 2015 (measured in kg of CO_2 equivalent per kilogram of product)

	Average household	Male average household	Female average household
Bread	114.7	99.6	100.2
Beer	59.2	75.1	41.1
Rice	35.1	34.0	31.1
Sugar	20.4	16.7	22.0
Vegetables, legumes and pulses	120.9	110.4	130.4
Other oils	30.0	16.1	24.7
Olive oil	92.7	92.7	108.8
Fruits and nuts	55.3	63.6	65.2
Wine	45.2	50.3	34.6
Coffee	64.4	55.5	68.5
Beef	360.7	267.8	380.8
Lamb and mutton	53.0	44.2	61.4
Pork	90.3	68.8	73.9
Poultry meat	103.7	75.9	96.2
Milk	428.2	360.0	430.6
Cheese	285.7	232.3	283.7
Eggs	66.1	61.2	65.3
Fish and shellfish	275.1	209.3	294.6
Sweeteners and honey	1.4	2.5	1.4
Butter and cream	29.5	22.7	29.3
Total	**2331.5**	**1958.7**	**2343.8**

Source Own elaboration based on Poore and Nemecek data (2018) and INE data (2015)

1664€ on restaurants (8% of their total expenditure). The average man's expenditure in restaurants is, therefore, twice that of a woman.

The following table provides a more detailed analysis of restaurant spending disaggregated by age group. Households were categorized into young adults (ages 18–30 years), middle-aged adults (ages 31–64 years) and older adults (aged 65 years and older) (Table 2).

Table 2 Annual expenditure on restaurants in 2015 (measured in euros)

Male average household			Female average household		
Young adult	Middle-aged adult	Older adult	Young adult	Middle-aged adult	Older adult
4507	3963	2413	3109	2872	760

Source Own elaboration from the data provided by the Household Budget Survey (INE 2015)

Young people spend the most in restaurants and then, the expenditure decreases with age. Moreover, for any age group, the average expenditure on eating out is much higher for men than for women. In other words, men are more likely to eat out while women are more likely to prepare food at home.

As mentioned above, there are differences in food expenditure derived from gender, so it is relevant to study the gap in each food product. Looking at some of the most frequent foods in the diet (and ignoring the rest of the foods), the following differences by gender and age groups can be found (Fig. 2):

Starting with alcoholic beverages (beer and wine), a man spends 83% more on beer and 46% more on wine than a woman. In addition, a man living alone consumes larger quantities than the average Spanish household. Regarding non-alcoholic beverages, a man drinks 16% less milk and 20% less coffee than a woman, at least in terms of expenditure at home.

As for bread and rice, consumption is quite similar for both genders and lower than that of an average household. As for sugar, a man consumes 25% less, while spending more on sweeteners and honey (70% more than a woman). Looking at oils, an average woman spends more on oils (any oil, including olive oil) than a man.

With regards to fruits, there are no differences in consumption by gender and a single-person household consumes larger quantities than the average Spanish household. For vegetables, a man spends 15% less than a woman.

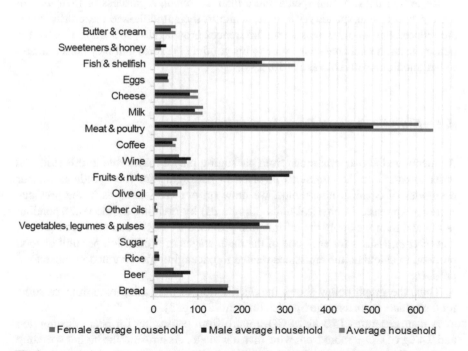

Fig. 2 Annual expenditure on selected food products in 2015 (measured in euros) *Source* Own elaboration from the data provided by the Household Budget Survey (INE 2015)

With respect to meat and fish, a man living alone consumes less than a woman (-30% fish and -17% meat) and less than the average member of a Spanish household. Besides, comparing a woman with the average household, it can be observed that meat consumption is lower (close to 6% less) and fish consumption is higher (approximately 7% more).

Differences are also observed in animal products: apart from milk (already analyzed), a man consumes 23% less butter and cream than a woman. Looking at eggs and cheese, a man spends 6% and 18% less than a woman, respectively.

The category for meat includes multiple products, but, if we limit ourselves to the four most consumed types of meat products and disaggregate the data by age group, we can observe the following:

We can draw the following conclusions from these data. Firstly, as age increases, there is a considerable rise in meat consumption in general and gender patterns also change. Second, there are gender differences according to the type of meat (bovine, lamb, pork, or poultry meat).

For beef, there is hardly any difference between a young man and a young woman. However, as age increases, a man consumes less than a woman (in the elderly, 26% less). As for lamb meat, for younger individuals, the average man consumes three times more than the woman; a middle-aged man consumes 29% more; and an elder man consumes 32% less.

Regarding pork, a man spends more than a woman, regardless of their age. In their youth, a man spends 65% more, while in their middle-age these differences are reduced and in older ages, these differences grow again (16% gap). Finally, for poultry meat, a man consumes more in his youth (7% more) and they consume less in her middle-age and old age (15% less).

4.3   Emissions from Food Packaging

As mentioned above, emissions from packaging play a major role in this study. In order to calculate the emissions involved in packaging for food products for our categories of Spanish households, we draw on our HBS data, as in the previous section, expressed in physical units (kilograms) by dividing by the corresponding prices and then we apply the information provided by Poore and Nemecek (2018), that offer emissions for each one of the main stages of production per unit of food product, to calculate the emissions related to packaging incorporated in those food products.

Thus, the graph below shows the distribution of packaging emissions of some food products in an average Spanish household (Fig. 3).

A man generates 15.3 kg of CO_2e more from packaging for beer consumption and 7.4 kg of CO_2e more from wine than a woman. As a result, the higher spending on alcoholic beverages translates into higher total emissions of food packaging (6% more than a woman). Moreover, a single-person household (male or female) generates

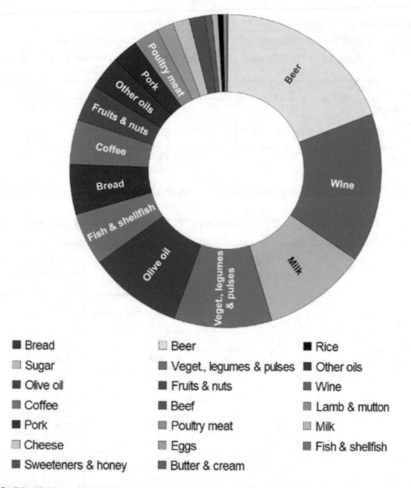

Fig. 3 Distribution of packaging emissions of food products generated by average Spanish household in 2015 (measured in kg of CO_2 equivalent per kilogram of product). *Source* Own elaboration based on Poore and Nemecek data (2018) and INE data (2015)

fewer emissions due to food packaging than the average Spanish household. Table 4 shows in more detail the results for each food product.

This is an interesting result, as it shows that, despite spending more on food to prepare at home, emissions related to packaging are lower for a woman living alone than for either a man living alone or the average person in Spain.

As we can see, analyzing together Tables 3 and 4, emissions from packaging are 5.96% of food emissions for the average household, 6.96% for the average single-person male household and 5.49% for the corresponding female household. This is a reflection of the differences in their pattern of food consumption, as emissions due to packaging of some products, like beverages, are considerably higher in terms of total emissions for those food products.

Table 3 Annual expenditure on meat in 2015 (measured in euros)

	Male average household			Female average household		
	Young adult	Middle-aged adult	Older adult	Young adult	Middle-aged adult	Older adult
Beef	34	58	78	34	64	107
Lamb and mutton	8	18	25	2	14	37
Pork	29	50	86	17	46	74
Poultry meat	56	75	91	53	89	108

Source Own elaboration from the data provided by the Household Budget Survey (INE 2015)

Table 4 Packaging emissions of food products in 2015 (measured in kg of CO_2 equivalent per kilogram of product)

	Food packaging emissions		
	Average household	Male average household	Female average household
Bread	7.2	6.2	6.3
Beer	26.5	33.7	18.4
Rice	0.8	0.7	0.7
Sugar	0.8	0.7	0.9
Vegetables, legumes and pulses	14.1	12.9	15.2
Other oils	4.8	2.6	3.9
Olive oil	13.4	13.4	15.8
Fruits and nuts	5.3	6.1	6.3
Wine	21.4	23.8	16.3
Coffee	6.4	5.5	6.8
Beef	2.3	1.7	2.4
Lamb and mutton	0.5	0.5	0.6
Pork	3.7	2.8	3.0
Poultry meat	3.6	2.6	3.3
Milk	14.9	12.6	15.0
Cheese	2.3	1.9	2.3
Eggs	2.4	2.2	2.3
Fish and shellfish	7.4	5.6	7.9
Sweeteners and honey	0.1	0.2	0.1
Butter and cream	1.0	0.8	1.0
Total	**138.9**	**136.4**	**128.7**

Source Own elaboration based on Poore and Nemecek data (2018) and INE data (2015)

We can then conclude that packaging emissions are not insignificant but that public opinion might overestimate the role of those emissions. In this sense, the importance of packaging can be more relevant in terms of waste and use of resources. We, therefore, analyze next the available information on packaging waste and recycling to provide some additional light on this topic.

4.4 Household Packaging Waste and Recycling

In the previous section, we analyzed, among many other things, the emissions of food products from packaging (that include the production of packaging material, transportation and end-of-life disposal). In this section, we focus on plastic material and its disposal at the end-of-life of the packaging.

All plastic production in the world amounted to 359 million tons in 2018 (335 in 2016) and 61.8 million tons in Europe (60 in 2016). More than half of all plastic is produced in Asia, with China representing 31% and Europe 16% and NAFTA 19% (PlasticsEurope 2021). Spain demands 7.8% of all European demand for plastic materials, that are later converted into plastic products (50.7 million tons of European demand of which 39.6% is for packaging) in 2018 (PlasticsEurope 2021). About 41% of these plastics are used for food products (8.2 million tons) (ING 2019a).

It is important to know the demand for plastics, but it is also important to know the waste that is generated and the disposal methods of plastics since it generates greenhouse gases.

The Ministry of Ecological Transition and Demographic Challenge (2021) reports over 7.5 million tons of packaging waste, of all material and uses (household, commercial and industrial), collected in Spain in 2018 (last available data from this source). Almost half of the material in terms of weight is paper and cardboard, while plastic represents almost 22% and glass, 19.5%. The differences in the ultimate destination of these types of waste are shown in Table 5. We can highlight that the highest percentage of recycled material corresponds to metals (84%), while the

Table 5 Distribution of total packaging waste by end-of-life destination (percentage)

%	Recycled (%)	Energy recovery and incineration (%)	Landfill (%)
Glass	76.8	0.0	23.2
Plastic	50.7	15.4	33.9
Paper and cardboard	72.5	3.6	23.9
Metals	84.0	0.4	15.6
Wood	67.0	13.5	19.5
Other	0.0	4.1	95.9
Total	68.8	5.7	25.5

Source Data from Ministry of Ecological Transition and Demographic Challenge (2021)

lowest is plastic (50.7%). Plastic is also the material with the highest share of packaging waste disposed of in landfills. This emphasizes the importance of considering plastics, not only because of their relevance for air and water pollution but also due to the significant amount of plastic packaging that is not recycled. As the minimum objective set by the EU for 2025 is 50%, Spain has already achieved the required level, but we are still short of the 55% objective for 2030.

Cicloplast (2019), an organization that encompasses all plastic firms in Spain (resin producers, converters and recycling and recovery firms) provides figures for plastic waste in the country. According to this organization, plastic waste, in general, amounted to 2,579,153 tons in Spain in 2018, continuously increasing from 2013 and 53.1% of that was household packaging (and additional 24.8% were retail and industrial packaging). Out of that total of plastic waste, 42% (1.08 million tons) was recycled, 19% was used for energy recovery and 39% ended up in a landfill in 2018 (that compares to 31.1%, 41.6% and 27.3% for EU 28 plus Norway and Switzerland in 2016, amounting to 8.4, 11.3 and 7.4 million tons according to PlasticsEurope 2018).

In total, 44.1% of all plastic products were applied to packaging, either for households, retail or industrial use, but it represented 64.4% of all plastic waste. As we indicated above, out of all (households, retail or industrial) packaging, 50.7% in 2018 were recycled, up from around 20% in the early 2000s. In terms of weight, recycled household plastic packaging was 571.9 tons, a substantial increase from 86 tons in 2001 and 323 in 2010.

We can show the figures for household packaging waste of different materials in per capita and per equivalent consumption unit, from Ecoembes (2021) and our previous HBS data (Table 6). According to data from Ecoembes (2021), 1.5 million tons of packaging from households (plastic, paper, metal and carton) were sent to recycling in 2020. Out of that total, 41.4% were plastic items, a share that has been increasing, from 35.4% in 2016. Each person in Spain recycled 12.3 kg of household plastic packaging in 2018, versus 6.9 in 2010 and 1.4 in 2000. This is top for Europe, as it is 11 for Germany, 10.3 for Italy, 9.1 for the UK or 4.4 for France (Cicloplast 2019). By 2020, that amount grew for Spain up to 13.1 kg per person (616,282 tons overall). In equivalent consumption units, the amount of collected plastic packaging waste was 19.96 kg. This represents an increase of almost 29% in the last five years.

We combine that information on the kg of household plastic packaging waste per equivalent consumption unit for Spain with data by Zheng and Suh (2019) and CIEL (2019), as commented above, in order to calculate the emissions embodied in that packaging and how the results are affected by its end-of-life treatment. In doing so, as we did when analyzing GHG emissions from food packaging, we are applying life cycle data, as provided by the sources we indicate.

According to Zheng and Suh (2019), producing the basic material for one kg of plastic generates between 3.6 and 1.9 kg CO_2e, depending on the type of polymer, calculated as a world average. For its conversion into manufacturing plastic products, the range is 0.8–1.4. By averaging according to their weight in global waste, we can therefore calculate that producing and converting one kg of plastic will involve releasing 3.95 kg CO_2e (2.63 + 1.32). We can compare these figures to those

Table 6 Household plastic packaging waste (kg per capita, kg per equivalent consumption unit and weight by end-of-life option)

	2016	2017	2018	2019	2020
kg per capita					
Plastic	10.29	11.23	12.31	13.28	13.27
Paper and cardboard	10.77	10.82	10.99	11.35	11.06
Metals (steel and aluminium)	5.63	5.73	5.60	5.37	5.37
Carton (brik)	2.24	2.18	2.21	2.25	2.23
Wood	0.18	0.17	0.18	0.17	0.16
Total	29.10	30.13	31.28	32.41	32.08
kg per ECU					
Plastic	15.49	16.90	18.53	19.98	19.96
Paper and cardboard	16.20	16.28	16.54	17.07	16.64
Metals (steel and aluminium)	8.47	8.62	8.42	8.08	8.07
Carton (brik)	3.37	3.29	3.32	3.39	3.36
Wood	0.27	0.26	0.27	0.25	0.24
Total	43.79	45.34	47.07	48.77	48.27
%					
Plastic	35.4%	37.3%	39.4%	41.0%	41.4%
Paper and cardboard	37.0%	35.9%	35.1%	35.0%	34.5%
Metals (steel and aluminium)	19.3%	19.0%	17.9%	16.6%	16.7%
Carton (brik)	7.7%	7.2%	7.0%	6.9%	7.0%
Wood	0.6%	0.6%	0.6%	0.5%	0.5%
Total	100.0%	100.0%	100.0%	100.0%	100.0%

Source Data from HBS (INE 2016) and Own elaboration based on Ministry of Ecological Transition and Demographic Challenge (2021)

calculated by CIEL (2019). This organization provides a figure of 1.89 kg, but it only includes cradle-to-resin stages (production and not conversion) and focuses on information by North American and European countries (this is particularly relevant in terms of the energy mix they consider). We will use both sources as defining a range of emissions from the plastic packaging collected from households. In order to make them more comparable, we will add the conversion emissions calculated in Zheng and Suh (1.32) to the production emissions by CIEL, to obtain 3.21 kg CO_2e per kg of plastic.

Equally important is to analyze the differences between both sources when it comes to emissions related to the end-of-life stage. Three potential destinations are presented in both studies: recycling, incineration and landfill. Zheng and Suh (2019) focus on emissions generated in those stages, but they do not include credits, that is "negative" emissions from the energy recovered in burning the plastic waste or "saved" emissions from substituted production of new plastic by recycled material.

CIEL (2019) provides emissions for those end-of-life options that include those recoveries. For landfills, both sources are quite similar, as Zheng and Suh calculate 0.089 kg CO_2e per kg of disposed plastic versus 0.060 for CIEL (2019). This is obviously an option with lower GHG emissions, but it is the least favoured option as most plastic has a long lifespan and landfills do not solve the problem of safely disposing of this type of waste.

For incineration, Zheng and Suh (2019) calculate that burning 1 kg of plastic waste generates 1.324 kg CO_2e, while CIEL (2019) offer a much higher figure of 2.967 (including transport and handling). This last source also includes energy recovery, which reduces then the impact to 0.897. As for recycling, Zheng and Suh indicate that this activity generates 0.906 kg CO_2e per kg of plastic waste. CIEL (2019) offer a slightly lower figure, 0.695 and, when it includes the savings in new material generation, the result is −1.395. This negative number indicates that recycling, according to this source of information, reduces GHG emissions, even taking into account that emissions are produced during the recycling process, as the pollution required to obtain new plastic is far greater.

Table 7 presents our results for household plastic packaging collected in Spain. We might start by noting that collected plastic is far lower than the total plastic used. It is difficult to estimate the share of plastic packaging that is not properly collected and should be added to the landfill total. Out of the plastic packaging collected from households, our calculations indicate that in 2020 (and assuming the most up-to-date distribution of recycling-incineration-landfill) each equivalent consumption unit generates between 93 and 80 kg CO_2e, figure that is reduced to 53 when we include energy savings from recycling. It is important to note that this amount would be increased if landfill disposal were to be substituted by incineration (59 kg CO_2e), so our plastic end-of-life distribution would look very similar to that of Germany. That would be the price to pay for avoiding the accumulation of plastic and potential leakages, or its export to other countries, as it currently happens with one third of plastic collected in Europe.

Table 7 GHG emissions from household plastic packaging waste (different scenarios, kg CO_2e)

	2016	2017	2018	2019	2020
Scenario 1: Current distribution of recycling, incineration and landfill					
Zheng and Suh (2019)	71.92	78.46	86.03	92.77	92.71
CIEL (2019)	62.58	68.27	74.86	80.73	80.67
CIEL recovery and recycling	41.22	44.97	49.31	53.17	53.13
Scenario 2: 50.7% recycled and 49.3% incineration					
CIEL recovery and recycling	45.61	49.75	54.56	58.83	58.79
Scenario 3: 55% recycled and 45% incineration					
CIEL recovery and recycling	44.08	48.09	52.73	56.86	56.82

Source Own elaboration from data by HBS (INE 2016), Ministry of Ecological Transition and Demographic Challenge (2021), Zheng and Suh (2019) and CIEL (2019)

Assuming we achieved the 55% objective for recycling (and leaving the rest to incineration), our results for the 2020 kg of collected plastic packaging would amount to less than 57 kg. Compared to the 139 kg CO_2e for all packaging (but restricted to food packaging) and 2332 kg CO_2e for food products that we estimated in the previous section, we can conclude that recycling, while indispensable both as another contribution to reducing air pollution and, most importantly, to decreasing land and water plastic litter, can only play a limited role in driving GHG emissions down.

Even assuming that we could recycle all plastic packaging and using the CIEL figures that indicate a difference of 1.395 kg CO_2e per kg of plastic between new and recycled plastic, that would imply reducing emissions per kg of plastic from 3.21 (without taking into account end-of-life emissions, of which the least CO_2-intensive is landfill, reaching then 3.27 kg CO_2) down to 1.815. That is an approximate reduction of 43% of the carbon footprint. In terms of the total number of recovered plastic from packaging and considering the current situation of the share of landfill, incineration and recycling, if we could recycle all plastic packaging, that would reduce the carbon footprint from plastic from 64–78 to 36 CO_2 kg per consumption unit. For a country like Spain, that could amount to a carbon footprint from recycled plastic of 1.1 million CO_2 tons per year, compared to 2–2.4 for new plastic, a total reduction of around to 1 million tons, when total CO_2 emissions in the Spanish economy are estimated around 270 million tons.

We must remember nevertheless that we have only included household plastic packaging, neither all types of packaging nor all sources of plastic and that the rise in plastic use means recycling will be increasingly important in years to come.

5 Conclusions and Discussion

In this piece of research, we have presented information regarding the carbon footprint of Spanish households, with particular attention to differences by gender and focusing on food products, using MRIO methodology. We have also looked at emissions from food packaging, combining our calculations with life cycle data from Poore and Nemecek (2018). Finally, we have tried to provide some estimates for emissions linked to household plastic packaging, using data from several official and public sources, as well as some previous life cycle studies.

In terms of carbon footprint, our results show fairly similar emissions for both female and male single-person households, with greater emissions for men. Focusing on food products, we can see that emissions from this consumption represent a very significant share of total households' emissions and differences arise depending on age and gender.

By gender, female single-person households spend more on food, but their male counterparts have greater hotel and restaurant expenditures. This is important in terms of carbon footprint, as emissions for those two categories of products and services are among the most important for households. It is also a reflection of cultural and social habits, as women are prone to cook for themselves (and others, in multi-person households), while men make more use of prepared meals. Policies directed towards reducing emissions should take into account these differences and understand incentives and measures like carbon taxes may impact agents in a distinctive way.

Meat is the most important source of pollution among food products, both due to the amount of expenditure that concentrates for Spanish households and for its relevant emission content. Dietary options, like veganism and vegetarianism, are minority in Spain. Depending on the source (there are no recent official surveys) and definition of vegetarianism, between 1.5 and 7.8% of the Spanish population are either vegetarian or flexitarian. Two thirds are estimated to be women, indicating a potential gender bias for emissions from food products. As we only consider in our calculations the money that households spend on food to be prepared at home, we find higher emissions from meat for females than for male single-living individuals. This points to the need to consider all categories of expenditure in food and meals before concluding that there is a gender bias in emissions associated with food products.

Even though women spend more money on food products, emissions related to food packaging are higher for men, as they consume a greater amount of alcoholic beverages. Bottles and cans incorporate significant emissions, compared to packaging for other products, like vegetables, that women are more inclined to consume. Incentives for recycling and changes in materials for packaging should consider their target consumer, depending on the products.

Our estimation of emissions from collected plastic packaging waste tries to provide some indication of their relevance compared to all emissions for households. While they are not the most concerning element in terms of air pollution, the combination of GHG emissions and water pollution makes recycling a relevant target for environmental policies, as identified by the EU strategies. Increasing the required percentage of recycled plastic (55% by 2030), the recycled content in plastic bottles (25% by 2025 and 30% by 2030), plastic collection (77% by 2025 and 90% of plastic bottles by 2029) and stopping the production of some single-use plastics from 2021, are all measures that should contribute as "economic and other incentives to support sustainable consumer choices and promote responsible consumer behaviour can be an effective tool for achieving" those objectives (Directive 2019/904).

These strategies are not only focused on reducing the amount of new, particularly single-use, plastic. They also enforce a harmonized marking for this type of plastic product to inform consumers and improve their appropriate disposal and collection. Furthermore, they also address the other end of the trade-in single-use plastic products, as they establish that producers of these items should cover the costs related to awareness policies, litter cleaning and data reporting.

Our calculations present a number of limitations. Some of them are related to the use of I-O methodology, as this implies using aggregated data at industry level for technical and emission coefficients and that are only available with some time lag. However, it ensures that all indirect embodied emissions are included in the carbon footprint estimates as it takes into account all inputs, inputs of inputs and regions of origin.

Similarly, when trying to calculate emissions for specific causes or materials, we are forced to draw on life cycle data to provide some estimates. This clearly introduces some noise, as sources of data apply different methodologies to reach those results, but we still consider those calculations as relevant despite their shortcomings.

Furthermore, in order to reach those results, we need however to combine data from several sources and correspond to different years. While we understand this is not ideal, we find it reasonable to expect that technical data (whether for the production or emissions) change slowly. Also, by using data in monetary and physical terms, we can avoid some of the biases that afflict some studies on carbon footprint and general calculation of emissions. For example, as we use monetary spending for carbon footprint calculation, prices for food products can distort to some extent the real environmental impact of consumption. That is why we find interesting the combination and comparison of life cycle data and I-O carbon footprint.

Plastic issues, yet with no clear solution, will persist. According to ING (2019a), population growth will lead to an increase in packaging, while at the same time it also predicts a rise in the use of packaging per person. Both trends will therefore aggravate the problem of related emissions and increase the need for recycling and waste treatment.

Besides emissions, our consumption impacts the environment in many ways, including the depletion of natural resources and waste generation. In this sense, up to 85% of marine litter, in terms of that found in EU beaches, is plastic and 50% of that total is a single-use plastic product (European Union 2019).

Multiple solutions to the consequences of plastic overuse have been elucidated. Reducing the amount of packaging in food, choosing non-plastics material or using corn- or sugarcane-based plastic and using more recycled material (if plastics are required) in packaging are the first steps to be taken. However, if 100% recyclability is to be achieved, the recycling process needs to be reconsidered. Improving waste collection and sorting systems is essential to raise the recycling rate of plastics. Particularly, the recycling rate of plastics increases tenfold when waste is collected separately (Plastics Europe 2020). Addressing the increasing demand for recycling facilities is another, as industries require recycled plastic to reach some characteristics to be safe, for example, for food packaging and at a reasonable price. In this sense, the increase in the price of CO_2 emissions, as we are currently witnessing, might act as a powerful incentive to further develop the recycling industry in the EU, together with explicit policies.

In terms of implications for policymaking, the analysis by Zheng and Suh (2019) shows the potential of combining multiple policies (energy, materials, recycling and demand management). Precisely, the implementation of renewable energy, 100% recycling and plastics demand reduction strategies (from 4 to 2%) have the potential for 2050 emissions to be brought back to the same level as 2015. In addition, replacing fossil fuel feedstock with biomass can further reduce emissions.

References

AEMET (2019) Informe del estado del clima en España en 2019. Agencia Estatal de Meterología, Gobierno de España. Available at http://www.aemet.es/es/conocermas/recursos_en_linea/public aciones_y_estudios/estudios/detalles/informe_clima_2019

Aguilera E, Piñero P, Infante-Amate J, Molina M, Lassaletta L, Sanz-Cobena A (2020) Emisiones de gases de efecto invernadero en el sistema agroalimentario y huella de carbono de la alimentación en España. Real Academia de Ingeniería

Arce G, Zafrilla JE, López LA, Tobarra MÁ (2017) Carbon footprint of human settlements in Spain. In: En Álvarez R et al (eds) Carbon footprint and the industrial lifecycle—from urban planning to recycling, green energy and technology. Springer, Berlin, pp 307–324

Cansino JM, Cardenete MA, Ordóñez M, Román R (2012) Economic analysis of greenhouse gas emissions in the Spanish economy. Renew Sustain Energy Rev 16(8):6032–6039

Carlsson-Kanyama A, Ekström M, Shanahan H (2003) Food and life cycle energy inputs: consequences of diet and ways to increase efficiency. Ecol Econ 44:293–307

Cazcarro I, Amores AF, Arto I, Kratena K (2020) Linking multisectoral economic models and consumption surveys for the European Union. Econ Syst Res. https://doi.org/10.1080/09535314. 2020.1856044

Cicloplast (2019) Cifras y datos clave de los plásticos y su reciclado en España. Datos 2018. Available at http://www.ciclopast.com/descargas/Cifras_plasticos_2018_WEB.pdf

CIEL (2019) Plastic and climate. The hidden costs of a plastic planet. Available at www.ciel.org/ plasticandclimate

Corsatea TD, Lindner S, Arto I, Román MV, Rueda-Cantuche JM, Velázquez Afonso A, Amores AF, Neuwahl F (2019) World input-output database environmental accounts. Update 2000–2016. Publications Office of the European Union, Luxembourg

Crippa M, Solazzo E, Guizzardi D, Monforti F, Tubiello F, Leip A (2021) Food systems are responsible for a third of global anthropogenic GHG emissions. Nature Food 2:1–12

Dirección General de la Mujer (2014) ¿Te cuidas? Hábitos de vida de las mujeres en la Comunidad de Madrid. Consejería de Políticas Sociales y Familia. Dirección General de la Mujer. Available in http://www.madrid.org/bvirtual/BVCM014004.pdf

Dormer A, Finn D, Ward P, Cullen J (2013) Carbon footprint analysis in plastics manufacturing. J Clean Prod 51:133–141

Druckman A, Jackson T (2010) An exploration into the carbon footprint of UK households. The Research Group on Lifestyles, Values and Environment (RESOLVE). University of Surrey. Available in https://resolve.sustainablelifestyles.ac.uk/sites/default/files/RESOLVE_WP_02-10. pdf

Druckman A, Jackson T (2016) Understanding households as drivers of carbon emissions. In: Druckman A, Clift R (eds) Taking stock of industrial ecology. Springer, Berlin, pp 181–203

Druckman A, Buck I, Hayward B, Jackson T (2012) Time, gender, and carbon: a study of the carbon implications of British adults' use of time. Ecol Econ 84:153–163

Ecoembes (2019) Las cifras del reciclaje. Ecoembes report. Available in: Estadísticas sobre el reciclaje de envases domésticos en España | Ecoembes

Ecoembes (2021) Estadísticas sobre el reciclaje de envases domésticos en España. Available at: https://www.ecoembes.com/es/ciudadanos/envases-y-proceso-reciclaje/datos-de-recicl aje-en-espana

European Union (2019) Directive (EU) 2019/904 of the European Parliament and of the Council of 5 June 2019 on the reduction of the impact of certain plastic products on the environment. Available at: https://eur-lex.europa.eu/eli/dir/2019/904/oj

European Union (2020) EU action to tackle the issue of plastic waste. Review No 04/2020. European Court of Auditors. Available in https://www.eca.europa.eu/en/Pages/DocItem.aspx?did=55223

FAO (2018) Sustainable food systems: concept and framework. Brief. Rome. Available in http:// www.fao.org/3/ca2079en/CA2079EN.pdf

FEMP (2014) Actualización, cálculo y registro de la huella de carbono municipal. Red Española de Ciudades por el Clima, Federación Española de Municipios y Provincias, Oficina Española de Cambio Climático, Ministerio de Agricultura y Pesca, Alimentación y Medio Ambiente. Available in http://www.redciudadesclima.es/actualizaci%C3%B3n-c%C3%A1lculo-y-registro-de-huellas-de-carbono-municipales

Gao Y, Li M, Meng B, Xue J (2020) The forces driving inequalities in China's household carbon footprints. Institute of Developing Economies [Discussion Paper]

GEAR (2020) Huella de carbono para AEEE en Toledo. Global Energy and Environmental Economics Analysis Research Group. Available in http://blog.uclm.es/grupogear/2020/02/07/co-organization-of-the-15th-conference-of-the-spanish-association-fro-energy-economics/

Geyer R, Jambeck J, Law K (2017) Production, use, and fate of all plastics ever made. Sci Adv 3

Gómez N, Cadarso MÁ, Monsalve F (2016) Carbon footprint of a university in a multiregional model: the case of the University of Castilla-La Mancha. J Clean Prod 138:119–130

Government of Spain (2019) Acuerdo de consejo de ministros por el que se aprueba la declaración del gobierno ante la emergencia climática y ambiental. Available in https://www.miteco.gob.es/es/prensa/declaracionemergenciaclimatica_tcm30-506551.pdf

Hoekstra AY, Wiedmann TO (2014) Humanity's unsustainable environmental footprint. Science 344:1114–1117

INE (2009) Gestión de clasificaciones estadísticas: catálogo de clasificaciones. Instituto Nacional de Estadística Español. Available in https://www.ine.es/gescla/pages/inicio.jsf

INE (2013) Contabilidad nacional anual de España: agregados por rama de actividad. Serie homogénea 1995–2012

INE (2015) Encuesta de presupuestos familiares (2006–2015): resultados con clasificación COICOP. Fichero de microdatos. Instituto Nacional de Estadística Español

INE (2016) Encuesta de presupuestos familiares (Households Budget Survey), 2015. https://www.ine.es

INE (2018) Contabilidad nacional anual de España: tablas de origen y destino. Base 2010 (SEC 2010). Instituto Nacional de Estadística Español

INE (2020) Cifras de población. Instituto Nacional de Estadística Español

ING (2019a) Plastic packaging in the food sector: six ways to tackle the plastic puzzle. ING Economics Department

ING (2019b) ING international survey. Consumer choices in the circular economy. November. Available at https://think.ing.com/uploads/reports/IIS_Circular_Economy_report_FINAL.PDF

IPCC (2021) Summary for policymakers. In: Masson-Delmotte V, Zhai P, Pirani A, Connors SL, Péan C, Berger S, Caud N, Chen Y, Goldfarb L, Gomis MI, Huang M, Leitzell K, Lonnoy E, Matthews JBR, Maycock TK, Waterfield T, Yelekçi O, Yu R, Zhou B (eds) Climate change 2021: the physical science basis. Contribution of Working Group I to the sixth assessment report of the Intergovernmental Panel on Climate Change. Cambridge University Press, Cambridge

IPSOS (2019) Barómetro social: Opinión sobre el medioambiente. Available in: https://www.ipsos.com/es-es/

Ivanova D, Vita G, Steen-Olsen K, Stadler K, Melo PC, Wood R, Hertwich EG (2017) Mapping the carbon footprint of EU regions. Environ Res Lett 12:054013

Lantern (2019) The green revolution. Lantern papers. Available in https://www.lantern.es/papers/the-green-revolution-2019

Leontief W, Ford D (1972) Air pollution and the economic structure: empirical results of input-output computations. In: Broady A, Carter AP (eds) Input-output techniques. North-Holland Publishing Company, Amsterdam

López LA, Arce G, Morenate M, Monsalve F (2016) Assessing the inequality of Spanish households through the carbon footprint: the 21st century great recession effect. J Ind Ecol 20(3):571–581

López LA, Zafrilla JE, Álvarez S (2017) La huella de carbono y el análisis input-output. Serie huella de carbono. Volumen 6. AENOR Internacional, España

Mahajan S, Beutel J, Guerrero S, Inomata S, Larsen S, Moyer BR, Remond-Tiedrez I, Rueda-Cantuche JM, Simpson LH, Thage BV, Rompaey CV, Verbiest P, DiMatteo I, Kolleritsch E,

Alsammak I, Brown G, Cadogan A, Elliot D, Amores AF, Ghanem Z, Lenzen M, Meng B, Mesnard LD, Moylan CE, Howells T, Oosterhaven J, Pedersen OG, Pereira X, Rodrigues JF, Sixta J, Stapel-Weber S, Temursho U, Yamano N, Aḥmad N, Smith HW, Chow J, Singh G, Sim B, Alfieri A (2018) Handbook on supply, use and input-output tables with extensions and applications. United Nations, Nueva York

Martínez J, Roca J (2013) Economía ecológica y política ambiental. Fondo de cultura económica, México

Mbow C, Rosenzweig C, Barioni LG, Benton TG, Herrero M, Krishnapillai M, Liwenga E, Pradhan P, Rivera-Ferre MG, Sapkota T, Tubiello FN, Xu Y (2019) Food security. In: Shukla PR, Skea J, Calvo Buendia E, Masson-Delmotte V, Pörtner H-O, Roberts DC, Zhai P, Slade R, Connors S, van Diemen R, Ferrat M, Haughey E, Luz S, Neogi S, Pathak M, Petzold J, Portugal Pereira J, Vyas P, Huntley E, Kissick K, Belkacemi M, Malley J (eds) Climate change and land: an IPCC special report on climate change, desertification, land degradation, sustainable land management, food security, and greenhouse gas fluxes in terrestrial ecosystems

Miller RE, Blair PD (2009) Input-output analysis: foundations and extensions. Cambridge University Press, Cambridge

Ministry of Agriculture, Fisheries and Food (2021) Panel de consumo alimentario. Series de datos de consumo alimentario en hogares. Gobierno de España

Ministry of Ecological Transition and Demographic Challenge (2021) Memoria anual de generación y gestión de residuos. https://www.miteco.gob.es/es/calidad-y-evaluacion-ambiental/publicaci ones/Memoria-anual-generacion-gestion-residuos.aspx

Modlinska K, Adamczyk D, Maison D, Pisula W (2020) Gender differences in attitudes to vegans/vegetarians and their food preferences, and their implications for promoting sustainable dietary patterns—a systematic review. Sustain MDPI Open Access J 12(16):1–17

Monsalve F (2017) MATLAB code to compute producer and consumer responsability in global emissions. Available at https://es.mathworks.com/matlabcentral/fileexchange/63015-computing-ghg-emissions-input-output

Nijdam D, Rood T, Westhoek H (2012) The price of protein: review of land use and carbon footprints from life cycle assessments of animal food products and their substitutes. Food Policy 37:760–770

OECD (2008) Household behaviour and the environment: reviewing the evidence. Organisation for Economic Co-operation and Development. Available in: https://www.oecd.org/environment/con sumption-innovation/42183878.pdf

Ortiz S, Marco R (2006) La medición estadística de la pobreza. Visión Net. Madrid

Pandey D, Agrawal M, Pandey JS (2011) Carbon footprint: current methods of estimation. Environ Monit Assess 178(1–4):135–160

PIR (2019) Panorama de los Recursos Globales 2019: Recursos naturales para el futuro que queremos. In: Oberle B, Bringezu S, Hatfeld-Dodds S, Hellweg S, Schandl H, Clement J, Cabernard L, Che N, Chen D, Droz-Georget H, Ekins P, FischerKowalski M, Flörke M, Frank S, Froemelt A, Geschke A, Haupt M, Havlik P, Hüfner R, Lenzen M, Lieber M, Liu B, Lu Y, Lutter S, Mehr J, Miatto A, Newth D, Oberschelp C, Obersteiner M, Pfster S, Piccoli E, Schaldach R, Schüngel J, Sonderegger T, Sudheshwar A, Tanikawa H, van der Voet E, Walker C, West J, Wang Z, Zhu B (eds) Un informe del Panel Internacional de Recursos. Programa de las Naciones Unidas para el Medio Ambiente. Nairobi, Kenia

PlasticsEurope (2018) Plastics—the facts 2018. An analysis of European plastics production, demand and waste data. Available at https://www.plasticseurope.org/application/files/6325/4510/ 9658/Plastics_the_facts_2018_AF_web.pdf

PlasticsEurope (2020) The circular economy for plastics—a European overview. Plastics Europe report

PlasticsEurope (2021) Plastics—the facts 2020. An analysis of European plastics production, demand and waste data. PlasticsEurope report

Poore J, Nemecek T (2018) Reducing food's environmental impacts through producers and consumers. Science 360(6392):987–992

Räty R, Carlsson-Kanyama A (2010) Energy consumption by gender in some European countries. Energy Policy 38(1):646–649

Red2Red Consultores (2014) La evaluación de impacto en función del género en el medio ambiente. Instituto Vasco de la Mujer. Gobierno Vasco. Available at https://www.euskadi.eus/contenidos/inf ormacion/ipp_adm_general/es_emakunde/adjuntos/materiales.sectoriales.medio.ambiente.pdf

Singh P, Wani AA, Langowski H-C (2017) Introduction: food packaging materials. In: Singh P, Wani AA, Langowski H-C (eds) Food packaging materials: testing & quality assurance. Taylor & Francis Group, London, UK, pp 1–9

Springmann M, Godfray HCJ, Rayner M, Scarborough P (2016) Analysis and valuation of the health and climate change cobenefits of global dietary change. Proc Natl Acad Sci 113(15):4146–4151

Temurshoev U (2013) MATLAB code: generalized RAS, matrix balancing/updating, biproportional method. Available in https://es.mathworks.com/matlabcentral/fileexchange/43231-generalized-ras-matrix-balancing-updating-biproportional-method

Timmer MP, Dietzenbacher E, Los B, Stehrer R, De Vries GJ (2015) An illustrated user guide to the world input-output database: the case of global automotive production. Rev Int Econ 23(3):575–605

Timmer MP, Los B, Stehrer R, De Vries GJ (2016) An anatomy of the global trade slowdown based on the WIOD 2016 release. Groningen Growth and Development Centre (GGDC) Research Memorandum 162. Available in: http://www.ggdc.net/publications/memorandum/gd162.pdf

Tobarra MA, López LA, Cadarso MA, Gómez N, Cazcarro I (2018) Is seasonal households' consumption good for the nexus carbon/water footprint? The Spanish fruits and vegetables case. Environ Sci Technol 52:12066–12077

Toro F, Serrano M, Guillen M (2019) Who pollutes more? Gender differences in consumptions patterns. Instituto de Investigación en Economía Aplicada Regional y Pública (Working paper)

United Nations (2009) System of national accounts—2008. European Communities, International Monetary Fund, Organisation for Economic Co-operation and Development, United Nations and World Bank. Available in https://unstats.un.org/unsd/nationalaccount/docs/sna2008.pdf

Verghese K, Lewis H, Lockrey S, Williams H (2019) Packaging's role in minimizing food loss and waste across the supply chain. Packag Technol Sci 28:603–620

WEF (2016) The new plastics economy: rethinking the future of plastics. Industry Agenda of World Economic Forum

Wiedmann T, Lenzen M (2018) Environmental and social footprints of international trade. Nat Geosci 11:314–321

WMO (2019) The global climate in 2015–2019. World Meteorological Organization, Ginebra

WMO (2020) Greenhouse gas bulletin (GHG bulletin)—No.16: the state of greenhouse gases in the atmosphere based on global observations through 2019

Zheng J, Suh S (2019) Strategies to reduce the global carbon footprint of plastics. Nat Clim Change 9:374–378

Zwijnenburg J (2016) Further enhancing the work on household distributional data-techniques for bridging gaps between micro and macro results and nowcasting methodologies for compiling more timely results. Paper prepared for the 34th IARIW general conference. Available in http://www.iariw.org/dresden/zwijnenburg.pdf

The Potential of Refuse-Derived Fuel Production in Reducing the Environmental Footprint of the Cement Industry

Gisele de Lorena Diniz Chaves, Renato Ribeiro Siman, Glaydston Mattos Ribeiro, and Ni-Bin Chang

Abstract The cement industry demands a considerable amount of energy and is responsible for significant greenhouse gas (GHG) emissions worldwide. As a result, the sector is under pressure for changes aimed at mitigating its environmental impact. Therefore, this sector is interested in adopting alternative fuels that minimize their emissions. Refuse-Derived Fuel (RDF) presents a great potential to replace fossil fuels in the cement industry and to reduce the non-recycled waste disposed in landfills. RDF is produced from the recycling of wastes with energy potential (significant calorific value) although they are not suitable for traditional recycling processes, either because they are dirty, contaminated, or due to their impeding logistics costs. In several developed countries, fossil fuels used in cement manufacturing were replaced by RDF, whose possibility is underused in Brazil. In this sense, this chapter demonstrates how RDF production from rejected waste can reduce the environmental footprint in the cement industry, mainly in emerging economies. A case study in a Brazilian reality demonstrates these potential benefits. It is possible to produce from 42,446.5 tonnes in 2024 with a small fuel replacement by cement industries, to 567,092.1 tonnes in 2040 if all non-recyclable waste available can be used to produce RDF. This recycling process and reverse logistics network avoided

G. de L. D. Chaves
Department of Engineering and Technology, Federal University of Espirito Santo—UFES, BR101–km 60, Sao Mateus, Brazil
e-mail: gisele.chaves@ufes.br

R. R. Siman
Department of Environmental Engineering, Federal University of Espirito Santo—UFES, Fernando Ferrari, 514, Vitoria, Brazil
e-mail: renato.siman@ufes.br

G. M. Ribeiro (✉)
Instituto Alberto Luiz Coimbra de Pós-Graduação e Pesquisa em Engenharia (COPPE), Federal University of Rio de Janeiro—UFES, Pedro Calmon, 550, Rio de Janeiro, Brazil
e-mail: glaydston@pet.coppe.ufrj.br

N.-B. Chang
University of Central Florida (UCF), 4000 Central Florida Blvd, Orlando, USA
e-mail: nchang@ucf.edu

© The Author(s), under exclusive license to Springer Nature Singapore Pte Ltd. 2022   35
S. S. Muthu (ed.), *Environmental Footprints of Recycled Products*,
Environmental Footprints and Eco-design of Products and Processes,
https://doi.org/10.1007/978-981-16-8426-5_2

annual disposal costs from \$3.9 million in the initial years to \$47.8 million in the year 2040. Despite the economic benefits, the GHG emissions (in CO_2 equivalent) reduced significantly: from 2217.04 tonnes in 2024 for 15% fuel replacement in the base scenario, to 11,208.69 tonnes in 2040 depending on the considered scenario. Even though describing the potential for Brazil, the benefits of RDF production both to minimize the environmental footprint of the cement industry but also to support rejected waste circular economy, can benefit other developing countries.

Keywords Environmental footprint · Refuse-derived fuel · Recycling · Cement production impact · Greenhouse gases · Fuel replacement · Developing countries

1 Introduction

The rapid urbanization worldwide boosted the demand for cement and cement-based materials. Its high consumption level raised concern as the cement industry is known worldwide for its intensive use of natural resources and high energy consumption (Scrivener et al. 2018; Habert et al. 2020). The sector is one of the most energy-intensive and, consequently, one of the most polluting (Stafford et al. 2016; Benhelal et al. 2021). An additional concern is the inefficient technologies used for cement production causing a huge release of CO_2 (Jokar and Mokhtar 2018). Cement manu-facturing is the third top CO_2 emitter industry in the world, behind only Power Plants and Iron and Steel production (Bataille 2020; Benhelal et al. 2021).

In cement production, the major proportion of emissions (50–60%) is caused by the decomposition of limestone to produce lime, while 30–40% of CO_2 emissions are generated as a result of combusting fossil fuels in pyro-processing units. Electricity consumption in the mills and air coolers complete the total emissions (Benhelal et al. 2021). However, the environmental impact can substantially differ according to the diverse manufacturing systems and distinct types of cement products, even if Portland cement, obtained from primary raw materials whose production is based on the combustion of vast quantities of fossil fuels, prevails in the market (Ammenberg et al. 2015). The main fuels employed for cement production are coal, oil, gas, and petroleum coke from oil refining (Stafford et al. 2016; Scrivener et al. 2018; Stripple et al. 2018), responsible for 5–8% of global CO_2 emissions (Kajaste and Hurme 2016; Benhelal et al. 2021).

Global cement production grew by over 77% between 2005 and 2019, from 2310 to 4100 million tonnes (CEMBUREAU 2019; IEA 2020). Considering its constant growth trend, CO_2 global emissions increased by more than three times from 1990 to 2014, reaching 2.083 billion tonnes (Benhelal et al. 2021). The direct CO_2 intensity of cement production increased by 0.3% per year from 2014 to 2017. In a business-as-usual scenario, i.e., if the current rate of CO_2 emissions is maintained, it is estimated to achieve 2.34 billion tonnes in 2050 (IEA 2018).

Without efforts to reduce global cement consumption, it is estimated a moderate annual growth rate until 2030. China leads the cement market with about 55% of global production, followed by India accounting for 8%. While it is expected a reduction in China production in the long term, other developing countries as India, other developing Asian nations, and African countries push cement demand with the expansion of their infrastructure (IEA 2020). Following the worldwide trend, mainly in developing countries, Brazilian cement production increased from 39.4 million tonnes in 2001 to 88.5 million tonnes in 2017, representing the world's 5th largest production in that year (EPE 2018) and the 8th in 2020 (EPE 2020).

According to the International Energy Agency—IEA (2020), public and private sector efforts and specific policies are enabling reductions in energy use and emissions in major cement-producing economies, as China and India, as well as European countries. However, further policy efforts in all countries will be required to achieve the necessary cement sector decarbonization. The technical feasibility of implementing mitigation strategies in the cement sector has been widely studied during recent decades but several relevant barriers compromise its implementation, as deeply discussed by Benhelal et al. (2021).

The high energy consumption rates, the fluctuating cost of fossil fuels, and the significant environmental impact have compelled the replacement of fossil fuels with renewable ones in the cement manufacturing process (Papanikola et al. 2019). Alternative fuels are one of the most promising approaches to restraint global Greenhouse gas (GHG) emissions in the cement industry (Benhelal et al. 2021). Fuels from residues such as waste oil, solvents, plastics, paper, scrap tires (ST), industrial and Municipal Solid Waste (MSW), refuse-derived fuel (RDF), agriculture biomass, meat, and bone meal have been studied as alternative fuels (Bourtsalas et al. 2018; Stripple et al. 2018; Papanikola et al. 2019). However, among the alternative fuels, RDF is the most used due to its high calorific value (Papanikola et al. 2019). RDF is a fuel obtained from solid wastes after treatment to segregate the materials with high calorific value (Infiesta et al. 2019; Ferdan et al. 2018). RDF is produced from waste and therefore can be considered as a recycled product. It is noteworthy that, in this case, the waste is not directly burned to generate energy, but rather undergoes different transformation processes to obtain the RDF which, in turn, can be used as fuel.

According to Stafford et al. (2016), the co-processing of alternative fuels provides cement industries with the possibility of reducing fossil fuel dependency and decreasing GHG emissions. Developed countries stimulated fossil fuels replacement by waste, mainly the Netherlands, but also Austria, Germany, and Norway. However, developing countries are responsible for a great share of cement production (CEMBUREAU 2019), and they depend on technology transfer to reduce GHG emissions from industrial facilities (IPCC 2014; Bataille 2020). Considering the potential of RDF to reuse non-recycled waste and its use as an alternative fuel in cement production, this chapter demonstrates how RDF production from rejected waste can reduce the environmental footprint in cement industry, mainly in emerging economies. A case study in a Brazilian reality demonstrates these potential benefits.

2 A Brief Overview of Cement Industry Environmental Challenge

The production process of Portland cement starts with the extraction of raw material, mainly limestone, that is milled and mixed with other substances, such as clay, iron, silicon, and aluminium. This mixture, called raw flour, is dispatched to rotary kilns and heated to promote a partial material melting that creates cement clinker granules. Clinker is obtained from the calcination process, where calcium carbonate decomposes and CO and CO_2 emissions are produced. This process is conducted under temperatures between 1200 and 1500 °C. After that, clinker is cooled, mixed with a small amount of other ingredients (gypsum and slag, for example), and milled (Lamas et al. 2013; Georgiopoulou and Lyberatos 2018; Stripple et al. 2018). Calcination is responsible for about 50% of the total CO_2 emissions from cement industry and the other large fraction of the remaining emissions hail from the fuels' combustion in the kiln (Georgiopoulou and Lyberatos 2018).

This process is responsible for GHG emissions, dust and particulate matter emissions, local scarcity of non-renewable resources, huge energy consumption, water use, and mercury emissions (Habert et al. 2020). Miller and Moore (2020) indicate that global concrete production contributes with approximately 7.8% of nitrogen oxide, 4.8% of sulfur oxide, 5.2% of particulate matter smaller than 10 μm, and 6.4% of particulate smaller than 2.5 μm emissions whose economic damages corresponds to 75% of the cement and concrete industry current value. Considering that no other material presents the same performance and simplicity of use at such a low cost (Scrivener et al. 2018), its substitution is not considered feasible in a short term (Habert et al. 2020), and the increasing global demand (IEA 2020), several studies have been conducted in this area aiming to provide solutions to the trade-off: mitigate the environmental impacts of cement manufacture while preserving its advantageous properties (Lamas et al. 2013; Rahman et al. 2015; Scrivener et al. 2018; Habert et al. 2020).

New Portland clinker-based cement alternatives, alkali-activated materials, use of calcined clay and engineered filler with dispersants, as well as the efficiency improvement of cement use are required to achieve the 2 °C scenario targets for 2050 (Miller et al. 2018). The IEA/WBCSD Roadmap (2009) evidenced that cement industry needs a reduction of 18% in its overall CO_2 emission by 2050 to support the target of 50% global emissions reduction goal to limit global warming under 2 °C of pre-industrial levels. To achieve it, the mitigation strategies for cement industry are alternative fuels, energy efficiency, clinker substitutes, and Carbon Capture and Storage (CCS). Despite CCS presenting the best long-term CO_2 emission mitigation, the cost to implement this option is the highest among the proposed options (Scrivener et al. 2018). There is a substantial difference between the pricing of CO_2 emissions and the costs of mitigation at cement manufacture (Rootzén and Johnsson 2017). Aiming to manage the costs of CO_2 abatement in the cement industry, Rootzén and Johnsson (2017) realized a material flow and value chain for the Nordic reality. The authors demonstrated that the price range expected for emissions allowances will not

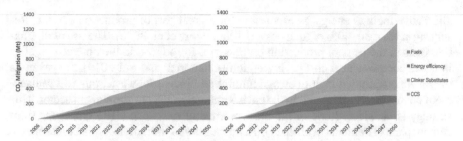

Fig. 1 Cement CO_2 emissions mitigation 2006–2050, for low demand (right) and high demand scenarios. *Source* Scrivener et al. (2018)

be sufficient to lead the CCS technologies development unless a regulatory scheme allocates part of the CO_2 abatement costs to the end-users. It is important to consider that pricing of carbon emissions usually boosts the diffusion of technologies like CCS but it is not sufficient to affect decarbonization in China's cement industry (Liu et al. 2017) (Fig. 1).

Jokar and Mokhtar (2018) evaluated the effects of three energy efficiency measures (clinker substitution, alternative fuel use, and waste heat recovery) on CO_2 emission, cost of energy consumption, manufacturer profit, and employed labor in the Iranian cement sector over 2015–2034. Reducing the clinker ratio by adding substitute materials has the greatest impact on CO_2 emission abatement, but government financial support in utilizing clinker substance has a decisive role in achieving sustainability (Jokar and Mokhtar 2018). Implementing existing and developing innovative-mitigation technologies demand substantial funding (Scrivener et al. 2018), as well as require time and governance. However, cement industries have slowly progressed in low GHG intensity options implementation as a result of some barriers: (i) the low-profit margins in especially competitive markets, (ii) capital costs are very focussed and upfront increasing the investment risk and conservatism, (iii) facility turnover is slow delaying more innovative plants, and (iv) emerging options to significantly reduce the GHG intensity are still not commercial due to the absence of a market for them (Bataille 2020).

An important challenge highlighted by Scrivener et al. (2018) is that a more equitable world, where developing countries require basic infrastructure related to housing, sanitation, and urbanization, will demand substantial cement production. Most new construction will occur in emerging economy countries (Bataille 2020). As the major amount of cement is and will be produced and consumed in developing countries, lower-cost alternatives should be provided (Scrivener et al. 2018). Regulation to trigger cement industrial decarbonization, as well as institutional supply chain awareness, education, motivation to change practices, are necessary to face these challenges, mainly in low-income countries (Bataille 2020).

Despite this wide source of environmental impacts and alternatives to minimize it, this section focus on the emissions related to the energy use in the cement production process as a real option in the short term. Cement industries are really sensitive to alternative fuels as energy is responsible for a great part of production costs, representing at least 30–40% of total costs (El-Salamony et al. 2020). The use of alternative fuels in the cement production process was introduced in the nineteen eighties in some countries as a cost-saving option (Chatterjee and Sui 2019). However, as they were constituted mainly by agricultural and industrial waste, this fuel replacement contributed to minimizing the environmental impact of the cement sector. This strategy is considered one of the most promising to reduce GHG emissions from cement plants (Benhelal et al. 2021). The benefits and limitations of these alternative fuels for the cement production process are discussed in the next section.

3   Co-benefits and Risks of Co-processing in Cement Industry

Co-processing refers to the use of waste as raw material and/or as an energy source in manufacturing processes. The waste can substitute natural mineral resources characterizing material recycling, but also it can replace fossil fuels in industrial processes representing an energy recovery (Lamas et al. 2013). Waste materials co-processed aiming for energy use are considered alternative fuels (Chatterjee and Sui 2019). They can replace coal, oil, gas, and petroleum coke from oil refining commonly used in cement manufacture (Stripple et al. 2018). This fuel replacement is possible thanks to considerable advances in collection, transportation, treatment, distribution, and supply of combustible waste materials in cement plants (Chatterjee and Sui 2019).

The process of clinker production realized in a rotary kiln at high temperatures provides favorable conditions for alternative fuels: alkaline environment, oxidizing atmosphere, large heat-exchange surface, and long residence time allows to burn a wide range of waste, and hazardous material (Mokrzycki and Uliasz-Bocheńczyk 2003; Lamas et al. 2013; Rahman et al. 2015). These characteristics of the cement kiln supported waste management since the massive quantity of waste generated from manufacturing, agriculture, and municipalities can be recycled as material supply or provide energy recovery (Lamas et al. 2013). Increasing fossil fuel prices motivates cement producers to promote the alternative fuel introduction in fuel mix. However, the economic incentives related to the volatility of the fossil fuels price is a limiting factor in the replacement development (Rahman et al. 2015).

According to Chatterjee and Sui (2019), the use of alternative fuels in cement production involves the industry size, its environmental impacts, and the alignment with waste management. In this sense, here it is important to consider the specificities of developed and developing countries when leading their waste management processes. The adaptation of a cement kiln to allow the use of waste material is cheaper than building a new incineration plant (Mokrzycki and Uliasz-Bocheńczyk 2003). However, incineration for waste treatment or energy supply is a practice related to developed countries. Incineration and anaerobic digestion, widely used in developed countries, are constrained by their high costs and complex operating requirements (Leal Filho et al. 2016). Landfilling and improper disposal is still the main disposal option in emerging economies (Lima et al. 2018) and waste-to-energy projects remain an underexploited option in countries like Brazil (Kaza et al. 2018; Dalmo et al. 2019).

Therefore, waste use in the cement industry brings benefits for waste management in high and low-income nations, even if these advantages rises from distinct motivations. The use of combustible waste material as an alternative fuel is considered by Ammenberg et al. (2015) an approach of industrial ecology since both waste management and cement industry can exchange to mutual advantage. This also contributes to more closed-loop supply chains minimizing the "ordinary cement production" critics related to the traditional Portland cement production based on a large share of non-renewable raw material flows (Ammenberg et al. 2015).

On the other hand, developed countries, as the USA and European state members, have a strong regulatory mechanism that imposes clear rules for the control of pollutant emissions in the cement production process, as well as established a credible compliance enforcement system. Developing countries face great challenges in supervising the established standards, which represents a risk for air pollution since cement plants is one of the most important sources for the release of potentially toxic elements (PTEs) and hazardous organic pollutants (Habert et al. 2020; Yang et al. 2021). Countries of Africa, South America, and Asia present high pollution index attributable to an increase in industrialization and urbanization, fuel sources, and poor regulation and enforcement (Karagulian et al. 2015). Some developing countries with cement kiln co-processing are not supported with pollution control standards for cement products and processes. Considering that hazardous waste can be used in co-processing and these countries possess fewer facilities for its treatment and disposal, the absence or limited regulation is a considerable risk (Huang et al. 2012).

The use of waste materials in cement industry meets resistance in developing countries, given the difficulties in implementing the waste hierarchy processes (Leal Filho et al. 2016) that prioritize waste minimization, reuse, recycling, and incineration with energy recovery and, lastly, landfilling (Pires and Martinho 2019; Chang et al. 2020). The fragile recycling market could not resist if part of the recyclables is diverted from the recycling sector to energy recovery in cement industries (Chaves et al. 2021a). On the other side, the huge usage of road transportation in Brazil and the long distances between the source of the fossil fuels and the cement plants can be minimized with alternative fuel, as they are generated close to them. In addition to the decrease in CO_2 emissions associated with fossil fuel replacement, this fuel substitution minimizes the impact of fuel transportation to supply cement industries in large countries as Brazil (Stafford et al. 2016).

The discussion of benefits and risks related to the use of alternative fuels in the cement production process is much broader and deeper than the points brought up in this section. This discussion does not intend to exhaust the subject. This discussion has been conducted for more than two decades by the literature, public managers, companies, and regulatory institutions. This brief presentation sought to list some potential and real contributions of co-processing, as well as to raise critical points that can compromise the use of waste materials in cement industry. In the next section, the potential of refuse-derived fuels is highlighted among the alternative fuels for the cement sector.

4 The Refuse-Derived Fuel Use in Cement Plants

A wide variety of hazardous and non-hazardous wastes are employed as alternative fuel: wood, paper and cardboard, textiles, plastics, rubber and tires, industrial sludge, municipal sewage sludge, animal meal and fats, agricultural waste (rice husk, for example), sewage sludge, oily waste, and others. Despite the vast possibility to use waste materials as alternative fuels, only those with acceptable calorific values should be considered (Georgiopoulou and Lyberatos 2018). To improve the waste calorific value, the industrial preparation process including separation of incombustible materials, moisture reduction, and downsizing can be realized (Rahman et al. 2015). The resulting product, called Refuse-Derived Fuel (RDF), presents a high calorific value, positive impacts on the performance of cement production, and limited environmental impacts (Bourtsalas et al. 2018).

The sorting and rejects removal are essential steps to ensure RDF production quality (Infiesta et al. 2019). Intended to cement industries, RDF production should include crushing and shredding phases safeguarding a reduction in particle size (Kumar et al. 2020). Chandrasekhar and Pandey (2020) recommended an RDF size inferior to 75 or 35 mm in some cases. For Garcés et al. (2016), the particle size must be less than 10 mm for the use in a cement kiln, but a 100 mm fuel is accepted for injection into precalciner. Chandrasekhar and Pandey (2020) recommended that moisture do not be superior to 25%, Ash inferior to 20%, and Chlorine level inferior to 1%.

Al-Salem (2019), consider the RDF a pretreatment used for incineration while IJgosse (2019), Haraguchi et al. (2019), and Chandrasekhar and Pandey (2020) consider the waste fuels for incineration and co-processing as different routes. Unlike incineration, waste consumption is complete in cement industry co-processing as no residual waste is produced in this combustion process (Chandrasekhar and Pandey 2020). RDF is fuel without non-combustible materials unlike in incineration the mass-burn uses all the Municipal Solid Waste (MSW) without prior treatment or preparation (Leme et al. 2014).

With distinct procedures depending on the input physical material characteristics, the final particle size, and the technology involved in its production, the composition and format of RDF varies (Chang and Chang 2001; Ferdan et al. 2018). It can be produced from the sorted dry combustible fraction, which cannot be recycled like cardboard drink containers, PE/PET bottles contaminated by PVC, packaging waste, rejects from manufacturing, scrap tires and, waste textiles (Rada and Andreottola 2012). Usually, this residual or rejected fraction, produced after an MSW treatment, can be profitable from the environmental and economic points of view (Gallardo et al. 2014). Other non-hazardous waste employed are from construction and demolition

(Garcés et al. 2016), sewage sludge or residues from car dismantling processes (Rada and Andreottola 2012), paper production plants (Sarc and Lorber 2013), or discarded biomass like rice husk (Souza et al. 2012) or food (Papanikola et al. 2019). However, RDF can also be produced from hazardous waste. It is available in fluff, pellet, bricks, or log forms and can be called by its waste component as tire-derived fuel (TDF), packaging-derived fuel (PDF), paper and plastic fraction (PPF), for example (Galvez-Martos and Schoenberger 2014).

Due to this variety in the materials supply for RDF production, quality requirements need to be satisfied to allow its efficient and safe use. Apart from legal requirements like limit values for heavy metal concentrations in waste fuels, Sarc and Lorber (2013) listed some targets that must be achieved: well defined calorific value, low chlorine content, quality controlled composition (few impurities), defined grain size depending on RDF application, defined bulk density and availability of sufficient material quantities to accomplish required specifications.

Several studies were dedicated to analyzing these factors for different sources of materials. Table 1 presents the elemental composition and combustion characteristics of different solid fuels usually used for cement production. Despite several possibilities, RDF achieves a combustion behavior closer than fossil fuel used by the cement industry. RDF has a higher calorific value than untreated MSW (Infiesta et al. 2019). Despite the heating value, Zhao et al. (2016) highlighted other advantages of RDF compared to raw MSW used as a fuel: more homogeneous physicochemical composition, ease storage, handling, and transportation, lower pollutant emissions, and reduced excess air requirement during its combustion.

RDF is worldwide used for energy recovery or as waste reduction before landfilling. Between Waste-to-Energy (WtE) options, RDF highlights between incineration, pyrolysis, and gasification (Reza et al. 2013; Garcés et al. 2016). Comparing these options, Haraguchi et al. (2019) demonstrated that RDF achieved the highest economic return and accomplished electricity demands when future policy scenarios were considered due to social costs of emissions and higher collection fees. Nevrlý et al. (2019) also demonstrated that the RDF alternative is more interesting in long term than other WtE alternatives using MSW attributable to the greater GHG savings. RDF use as co-fuel in cement kilns compared to other combustion processes have some advantages: it is a zero-waste method (Garcés et al. 2016), it reduces the conventional fossil fuels use with simultaneous material recovery (Samolada and Zabaniotou 2014), residual ash is effectively incorporated in the cement product (Garcés et al. 2016), and the oxidation of toxic pollutants are complete (Samolada and Zabaniotou 2014).

Hashem et al. (2019) analyzed the possibility of using rubber and plastic wastes for RDF partial replacement of traditional fossil fuel used in the cement industry and the properties of the produced cement. They concluded that its use could provide good results as a reduction in setting times, an improvement in the compressive strength, and acceleration in the hydration reactions. Reza et al. (2013) evidenced the environmental and economic feasibility of RDF production and utilization as an alternative fuel in cement plants with positive financial benefits for municipalities, consumers, and society in Canadá.

Table 1 Elemental composition and combustion characteristics of different solid fuels

Fuel type	Moisture	Ash	C	H	N	S	Cl	O	LHV[a]	HHV[b]
Wood (pine/waste)	5.6–6.3	0.5–8.8	46–51.3	5.7–5.8	0.07–3.8	0.01–0.05	~0	35.4–36	14.7–22.5	16.0–23.9
RDF from MSW	3.7–20	10.2–13.8	41.7–50.2	4.4–7.8	0.75–1.65	0.1–0.76	0.7–1.13	28.5–36.3	13.7–17	15.1–18.4
Tires	0.7–4	2.2–8.2	76.7–89.4	7–7.8	0.2–0.5	0.8–2.2	~0–0.1	0.4–0.5	28.4–35.6	29.8–37.3
TDF[c]	0.9–1.9	3.3–4.4	83.8–86.7	6.9	0.3–0.6	1.9–2	~0	0.9–2.3	27.4–31.7	28.9–38.4
Pecan shells	14.6	3.32	46.84	5.41	0.44	n/a	n/a	n/a	15.7–18.2	17.1–19.8
Meat and bone meal	1.4–8.1	10.4–28.3	42.1–55.7	5.8–8	7.2–8.9	0.05–0.4	0.2	15.3–38.4	16.9–28.8	18.1–30.6
PE[d]	~0–0.17	~0–0.06	86	14	~0	~0	~0	~0	38.4–44.6	38.7–44.9
PVC[e]	~0–0.2	1–7.6	35.9–38	4.4–5	0–0.11	~0	57	~0	14.7–17.1	15.6–18.1
Sewage sludge	5.2–5.6	17.9–29.5	36.4–40.5	4.7–7	0.84–5.0	0.1–0.6	~0–1	22	8.0–15.8	9.0–17.5
Sugarcane bagasse	9.2	2.9	47.8	5.9	0.5	n/a	n/a	45.7	14.9	16.4
Rice husk	9.2	18	37.4	5.4	0.4	n/a	n/a	33.2	n/a	14.4
MSW	n/a	4.4	51.2	6.2	0.1	n/a	n/a	40.1	17.2	n/a
Packaging[f]	0.18–5.84	0.81–6.59	67.8–81.7	7.84–15.1	~0–0.3	0.24–0.59	0.07–13.4	3.95–13.3	21.19–26.46	23.67–29.8
Soft plastics	1.19–1.69	1.89–7.03	67.6–79.8	9.95–14	0.31–1.55	0.59–0.93	0.08–0.31	4.81–9.24	37.50–41.7	40.32–45.17
Cellulosic (paper)	7.55–10.9	12.9–27.5	31.3–39.3	0.26–5.63	0.48–1.36	0.13–1.05	~0–0.16	29–56.8	14.03–16.5	16.57–18.41
Textile	1.50–4.19	4.90–12	52.4–61.2	3.64–7.13	0.64–4.11	0.46–1.84	0.06–7.93	22.1–34.2	20.8–25.48	22.09–28.03

(continued)

Table 1 (continued)

Fuel type	Moisture	Ash	C	H	N	S	Cl	O	LHV[a]	HHV[b]
Packaging global[g]	3.59–4.98	7.13–9.35	61.7–65	8.03–8.79	0.49–1.55	0.51–0.94	0.10–6.28	16.62–19.89	24.91–25.81	~28.81
Charcoal	4.6	4.7	72.2	2.9	1.3	n/a	n/a	23.6	n/a	26
Pet coke	0.8–1.5	0.5–1	89.5–92.7	2.4–3.7	1.2–1.7	1.5–4	~0	1.1–1.2	28.9–35.4	29.6–36.2
Coal	1.1–3.3	6.4–15.5	65.3–80.9	3.7–5.1	1.2–1.41	0.6–5.5	~0–0.33	5.9–12.5	21.9–31.8	22.7–32.9

Notes [a] *HHV* High Heating Value; [b] *LHV* Low Heating Value; [c] Tire-derived fuel; [d] Poly-Ethylene; [e] Poly-Vinyl-Chloride; [f] Non-identified material from packaging waste (see Garcés et al. 2016); [g] Mix of packaging, soft plastics, paper and textile as described in Garcés et al. (2016). n/a means not available
Source Elaborated by the authors based on Garcés et al. (2016), Bourtsalas et al. (2018), and Infiesta et al. (2019)

Despite the advantages, some challenges remain in RDF market development. RDF manufacture has a relatively high cost (Zhao et al. 2016) and RDF producers cannot rely on steady market conditions since the costs vary depending on either the availability of industrial plants for co-processing or the RDF landfilling option (Rada and Andreottola 2012). In addition, different technologies to obtain RDF are strongly dependent on the existence of end markets, mainly by cement industries (Papageorgiou et al. 2009). However, as RDF use in the cement industry is cost-oriented, it competes with fossil fuel prices (Nejati and Roknizadeh 2014). From the perspective of waste disposal, the reverse logistics network must have economic incentives to discourage landfilling (Paolo and Paola 2015), mainly in developing countries whose disposal is a cheap option.

RDF quality is an important issue to support fuel replacement. According to Chandrasekhar and Pandey (2020), due to the lack of RDF quality, its use in Indian cement plants has small significance: the thermal substitution rate is around 3–4%. Due to the variety of the RDF constitution materials, the atmospheric emissions risk exists (Gallardo et al. 2014), despite the existing technology for adequate control. The RDF chlorine level could be responsible for both technical problems (Garcés et al. 2016) and environmental concerns related to the formation of dioxins and furans during combustion in cement industries (Rahman et al. 2015; Bourtsalas et al. 2018). The pollutant emissions and clinker quality were provided by Rahman et al. (2015) and presented in Table 2. Used tires are a good option because their calorific value is high, their impact on clinker quality is not significant, with a low installation cost, high availability, and few changes in emissions. Plastic waste is also a good option, but with more heavy metals emissions and a moderate impact on clinker quality. Garcés et al. (2016) indicated that packaging materials (with soft plastics, paper, textile, and other unidentified materials) are a good waste source for RDF with emissions above legislation and high availability. Bourtsalas et al. (2018) also presented that non-recycled paper and plastic present benefits in cement co-processing, better than its use on WtE processes and also with emissions below law requirements. Besides, the greater the proportion of fuel substitution, the greater the risk of exceeding the legal limits set for polluting gas emissions (Lamas et al. 2013).

RDF is an economically viable option to minimize process energy costs and reduce GHG emissions in the cement industry (Benhelal et al. 2013; Reza et al. 2013) given regulatory requirements (Sarc and Lorber 2013; Gallardo et al. 2014; Benhelal et al. 2021). Although RDF can be used for electricity generation, Brazil gets around 45% of its internal energy supply and about 83% of its electricity matrix from renewable sources, and electricity is cheaper than in other developing countries (EPE 2019). Meanwhile, the cement industry is highly dependent on fossil fuels and may represent an important market for RDF in the country. RDF use in Brazil would provide more benefits if targeted to co-processing (Stafford et al. 2016; Lima et al. 2018). The potential for RDF in Brazil are presented in the next section.

Table 2 Comparison between alternative fuels for cement industries

Criteria	Alternative fuels						
	Used tire	MSW	MBM	Dried sewage sludge	Biomass (rice husk, wheat residue)	Plastic waste	Used oils and spent solvent
Calorific value (MJ/kg)	35.6	15.4	14.47	15.28	14–21	29–40	43–45
Moisture content %	0.62	31.2	6	Variable	6–12	Variable	<3
Availability	High	High	Moderate	High	Moderate	High	Moderate
NO_x emission	Unchanged	Reduced	Reduced	Reduced	Reduced	Unchanged	Reduced
CO_2 emission	Reduced 10%	Reduced −1.61 kg/kg RDF	Reduced 12%	Reduced −0.88 t/t coal replaced	Reduced	Reduced 15%	Reduced −2.02 t/t coal replaced
SO_2 emission	Increase	Increased	Reduced	Increased	Reduced	Reduced	ND[a]
Heavy metal emission	Reduced	Increased	ND[a]	Unchanged	Reduced	Increased	Reduced
Maximum substitution rate	30%	30%	40%	5%	20%	ND[a]	ND[a]
Storage requirement	Moderate	Moderate	Moderate	High	Low	Moderate	Moderate
Effects on clinker quality	Unchanged	Low	Low	Low	ND[a]	Moderate	Low
Installation cost	Low	High	Moderate	High	Low	Moderate	Low

[a] *ND* Not detected
Source Rahman et al. (2015)

5 The Refuse-Derived Fuel Potential: A Case Study in Brazil

Aiming to boost the institutional environment to WtE, Brazil is structuring the necessary legal framework. In 2019, two regulations related to energy recovery from waste were enacted. The first law regulates the MSW energy recovery (Brazil 2019a) while the second one designates the necessary elements of MSW's energy recovery projects to be qualified for the Federal Investment Partnership Program (Brazil 2019b). These regulations describe both the procedures for WtE projects but also designate financial incentives to projects that gather both technical and environmental requirements. In the same way, a national technical standard determines the MSW requirements for energy purposes, together with the limits of chlorine and mercury (ABNT 2020). Considered a WtE initiative, a pioneering regulation describing the technical standard requirements and incentives to promote RDF manufacture and use was developed in São Paulo (SIMA 2019). After that, the Ministry of the Environment (Brazil 2020) decreed the licensing requirements for co-processing in clinker rotary kilns. Waste can be used as fuel substitutes, as long as the energy and environmental gains are proven. Licensing is not required for sorted waste originated from the screening process of waste pickers' cooperatives which the recyclable waste fraction was already separated.

In Brazil, the cement industry consumed 3.97 million tonnes of fuel in 2018, mainly fossil fuels (more than 73%), according to the Energy Research Company (EPE 2020). This industry consumes around 10–12% of the total energy in all industrial sectors and around 1.5–2% of the total energy used in Brazil (EPE 2014, 2019). Despite efficiency gains with the modernization of the sector (around -1.8% per year) and the reduction in specific energy consumption (reduction of 11.4% between 1990 and 2012), total energy consumption increased in the last two decades due to the substitution of oil and coal derivatives with cheaper petroleum sources, mainly petroleum coke, and other alternative sources with lower calorific value, which justifies the consumption growth (EPE 2014). With this fuel shift, petroleum coke became the main fuel in cement sector, as demonstrated in Fig. 2. Of the total co-processed waste used in Brazilian cement industry, only 8% were originated from waste (Gomes et al. 2019). The main waste material employed was scrap tires.

The Brazilian cement industry is considered energetically efficient, but alternative fuels are still underused (Stafford et al. 2016; Gomes et al. 2019). In 2017, 1172 tonnes of materials were co-processed, with a thermal substitution rate of 11.9% in total (ABCP 2019). However, this index is still far from the replacement ratio achieved by developed countries, such as the Netherlands (83%), Austria (63%), or Germany (62%). Considering the average use of 30% of alternative fuels in the European Union (EU) (Bourtsalas et al. 2018), the Brazilian substitution process could be enhanced. The alternative fuels employed in co-processing in Brazilian cement industries are mostly waste tires, followed by waste blend and biomass (ABCP 2019).

In this sense, Chaves et al. (2021a) estimated the availability of possible recyclable waste streams from municipal and industrial sectors for RDF production in

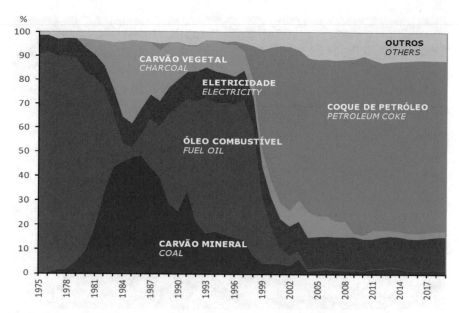

Fig. 2 Energy sources for cement industry in Brazil. *Source* EPE (2020)

Brazil. A system dynamics model was proposed and simulated based on a 20-year timeframe under waste management policies in Espírito Santo, a Brazilian state located in the southeast region of the country, as shown in Fig. 3. The annual availability of wastes originated from four different routes was obtained considering the uncertainties through the waste policies implementation, i.e., the authors considered pessimistic, intermediate, and optimistic scenarios. Considering the uncertainty to implement SWMP, an optimistic scenario considers that all goals and requirements established by the Espírito Santo waste policy would be accomplished by 2040, providing more waste for RDF production. While the intermediate scenario considers that uncertainties on its implementation would partially affect the achievement of goals. Given the demographic changes and economic development in the next 20 years, the authors obtained the waste availability for RDF production from 2020 to 2040. Focusing on rejection from the sorting process, i.e., non-recyclable paper, plastics, and scrap tires, the authors identified that if the waste management policy is partially implemented in Espírito Santo (intermediate scenario), by the year 2022 it will already be possible to obtain enough waste for the RDF minimum production. This means that there would be sufficient waste streams for RDF production under dynamic uncertainties given the possible cost–benefit-risk assessment. The authors evidenced that during 20 years, at least 3.0 million tonnes of non-recyclable paper and plastics in the intermediate scenario and 6.6 million tonnes in the optimistic scenario could be available for RDF, while more than 5.0 million tonnes of scrap tires in the intermediate scenario and 9.4 million tonnes in the optimistic scenario would also be recycled in RDF products (Chaves et al. 2021a).

Fig. 3 Location of the three cement industries at the Espírito Santo state, in Brazil

Based on the potential of waste recovery for RDF production and its use as an alternative fuel in cement industries, a reverse logistics network was proposed by Chaves et al. (2021b), as demonstrated in Fig. 4. The proposed reverse logistics network analysis considered a real case in the Espírito Santo state, in Brazil, in which three types of rejected wastes provided by 68 waste pickers' cooperatives, 37 waste retailers, scrap tire collection points present in 10 cities, and industries located in 78 cities, all of which are transported to candidate locations of RDF plants to meet the demands at three existing cement production plants. These three cement industries produced 753.85 million tonnes of cement in 2018 (SNIC 2020), and belong to Nassau Group, Polimix/Mizu Group, and Holcim Group.

The policy targets were assessed every 8 years: in 2024, 2032, and 2040, verifying the influence of the time-varying volume of waste on the decision to commission these RDF plants in terms of location and capacity. Scenarios with 15%, 30%, and 50% of fuel replacement were also simulated. A mixed-integer linear programming model was developed aiming to minimize the total costs for sending materials between cooperatives, waste dealers, scrap tire collection points, industries, and RDF plants, the transport cost from the opened RDF plants to cement industries, as well as, the installation and operation costs.

The results evidenced that all waste available is used to meet the demand of RDF plants for the intermediate condition in 2024. This also occurs in the year 2032 for the intermediate condition, when the cement plants reach 50% fuel replacement, and in the year 2024 for the optimistic condition with 30% and 50% fuel replacement, respectively, as shown in Fig. 5. In these cases, the supply of waste is the bottleneck

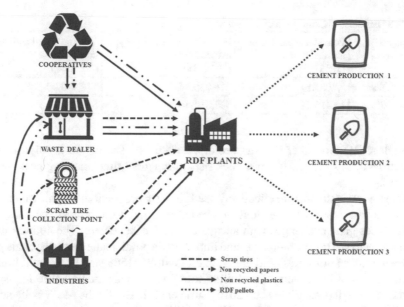

Fig. 4 Reverse logistics network for RDF production and consumption. *Source* Adapted from Chaves et al. (2021b)

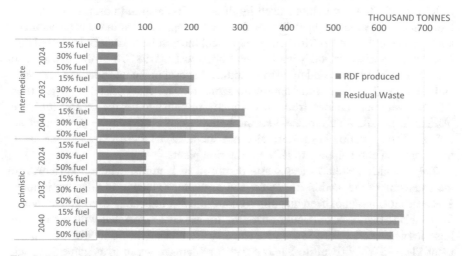

Fig. 5 RDF produced, residual waste, and cement industry demand fulfilled

to guarantee the complete fulfillment of the cement sector demand. The greater the replacement of fossil fuel in the cement industries, the greater the demand and production of RDF. In other scenarios, all the demand from the three cement plants is fully met by the sufficient supply of RDF. Residual waste means all waste that is not necessary to fulfill cement sector demand. However, this waste can be used

Table 3 Expected disposal avoided costs with RDF production

Scenario	Potential urban waste avoided cost (10^3 US$)	Potential industrial waste avoided cost (10^3 US$)	Total potential avoided cost (10^3 US$)
Intermediate	2063.14	276,066.86	278,130.00
Optimistic	15,418.68	543,834.56	559,253.24

to produce RDF for other industry sectors, but this market diagnosis surpasses the objectives of this study. This overstock rises throughout time attaining peak values in 2040.

If these waste streams are effectively used for RDF production, a significant parcel of combustible rejected waste would not be landfilled. It is important to consider that Brazil faces great challenges with adequate waste management. The country deals with a growth waste generation rate and improper disposal. Inadequate disposals such as open dumps (1493 sites) and unregulated landfills (1508 sites) were still present in all regions of the country. Landfilling remains the mainstream waste managing method in Brazil (Leal Filho et al. 2016; Lima et al. 2018). Of the MSW collected in 2018, 59.5% was sent to landfills, while 6.3 million tonnes of waste was not collected and, consequently, were improperly disposed (ABRELPE 2018).

The avoided costs related to waste deviation from disposal in sanitary landfills to RDF production are demonstrated in Table 3. The disposal costs for urban and industrial waste were obtained directly from the companies that manage the sanitary landfills in the state of Espírito Santo and reflect the cost in 2019. It is estimated that in the 20 years of waste policy implementation, a cost of US$ 278.1 million would be avoided related to urban and industrial landfill disposal in the intermediate scenario, and US$ 559.3 million in the optimistic scenario. The costs related to scrap tires disposal were not obtained from the company responsible for the reverse logistics of tires in Brazil, which means that the benefits can be expanded (Chaves et al. 2021a). Transportation costs were also not included, as there would be the same transportation cost shipping to RDF production plants.

This scenario results in less waste disposed in landfills, which contributes to the extension of the landfill life cycle, and a greater reduction in GHG emissions. These benefits have not been quantified economically, indicating even greater benefits in this scenario, which could offset the higher cost associated with the reverse logistics network. Espírito Santo state recently established the Solid Waste Management Plan—SWMP (Espírito Santo 2019) for adequate waste management, which included specific guidelines to stimulate WtE practices. Hence, these avoided costs could be redirected as economic incentives to RDF production. In this sense, the government should maintain a strong position to face the pressures that will certainly arise from sanitary landfills private companies that will have their revenues reduced with the diversion of waste. Avoidance of costs relegated to the government means less profit for private companies, which may result in resistance to RDF development via lobbying (Chaves et al. 2021a).

6 Environmental Footprint Discussion

Earth's limited natural resources and human exploitation are topics discussed since the 1970s and whose cause-and-effect relationship, reinforcing a cycle of global warming, characterizes a new era called by some authors the Anthropocene (Chaves et al. 2014; Matuštík and Kočí 2020). The environmental footprint is an indicator of human pressure on the environment (Hoekstra and Wiedmann 2014). Amid a sort of possible measures, the concentration of carbon dioxide in the atmosphere is one of the most popular environmental footprint indicators (Matuštík and Kočí 2020). Carbon dioxide (CO_2) equivalent is one of the main indicators of GHGs emissions responsible for climate change (Reza et al. 2013). To analyze the environmental footprint of RDF production from rejected waste for fuel replacement in the cement industry, the CO_2 equivalent was considered based on Chaves et al. (2021b).

The CO_2 equivalent in the RDF reverse logistics network was calculated considering the flows of non-recycled paper, plastics, and scrap tires from waste pickers' cooperatives, waste dealers, industries, and scrap tire collection points to RDF plants, together with the flow of RDF supply to cement plants in Espirito Santo case study. These flows were previously represented in Fig. 4 and represent the produced emissions with the RDF network, i.e., a new source of carbon dioxide introduced with this proposal. However, the waste used for RDF production would be diverted from landfill disposal, minimizing the emissions with waste transport from origins to the sanitary landfill. The transport of petroleum coke, the major fuel used in Brazilian cement plants (ABCP 2019), would be reduced with fuel replacement. The amount obtained was compared to the CO_2 equivalent related to waste shipping to landfills, fuel transported to cement plants, and the avoided emissions from fossil fuel replacement in cement production. Genon and Brizio (2008) indicated that the combustion of RDF in cement industries enables a decrease of 1.61 kg of CO_2 equivalent per kg of RDF used as a substitute for conventional fossil fuels. This parameter was also applied in other studies in Brazil (CETESB 2018) of RDF use in cement production. For the amount of non-recycled waste used for RDF production in each scenario, the avoided CO_2 emissions are linked to waste disposal, industry fuel supply, and the replacement of fossil fuels in cement production.

The transport flows involved the transport distances between the 78 municipalities of Espírito Santo, as well as the petroleum coke supply from the Rio de Janeiro state. Given the spatial locations of each municipality provided by DER-ES (2019), Euclidean distances were obtained between every source–destination pair and subsequently corrected by +10%, according to Ferri et al. (2015). Considering that road transportation is the main cargo transport mode in Brazil, this mode releases 0.11917 kg CO_2/tonne per kilometer using diesel fuel (PBMC 2013; Bartholomeu et al. 2016). Considering the carbon dioxide discharged from the RDF production network proposed for the Espirito Santo case study, the emissions avoided with the transport of waste to the landfills were discounted, as well as the transport of petroleum coke from the supplier located at a distance superior to 400 km from the cement industries operating in Espírito Santo state. In addition, the balance of the carbon footprint considered the reduction of CO_2 equivalent emissions with the combustion of RDF in the cement process.

The environmental footprint (CO_2 equivalent) with the RDF production and fossil fuel replacement in cement industries considering the case studied is demonstrated in Table 4. It is possible to observe the reduction of CO_2 equivalent emissions throughout the time and the distinct fuel substitution rates, which means more RDF produced. To assess the possibility of processing excess waste flows, the additional demand scenario verified the impact on the proposed network if other industries are interested in using RDF. Considering that cement industries consume all RDF produced, no RDF is available for additional demand in 2024, except if the fuel replacement is minimal in an optimistic scenario. This also occurs in the year 2032 for the intermediate condition. There is a reduction in CO_2 equivalent discharged that varies between 2217.04 tonnes in 2024 for 15% fuel replacement in the base scenario, to 11,208.69 tonnes in 2040 for 50% fuel replacement in the additional demand scenario.

From the elements considered and highlighted in Table 4, the main impact was on the fossil fuel supply. It is important to emphasize the massive use of road transportation in Brazil and the large distances between the source of fossil fuels and the cement industries. The transport sector in Brazil is the main emitter of GHG from energy use in Brazil (EPE 2018). Thus, RDF fuel replacement could also reduce the impact of emissions from transportation, corroborating with other realities, as indicated by Stafford et al. (2016). The emissions avoided from the disposal are underestimated since it is considered only waste transportation. The landfill emissions in Brazil involve the process of disposing of a wide variety of wastes, such as non-sorted and sorted MSW, the vast sort of industrial wastes, as well as residue from business offices, retail, shops, among others. It was not possible to estimate the landfill emissions for the non-recycled paper, plastics, and scrap tires separately. The least impact was on avoided carbon dioxide emissions as a result of fossil fuel replacement by RDF. It should be considered that this value indicated in Table 4 considers the balance between the emissions from the production of RDF and fossil fuel substitution in cement industries.

Figure 6 evidences a balance (grey bar) of the emissions of CO_2 produced (blue bar) by the reverse logistics proposed and the total avoided (orange bar) for the case provided considering the intermediate (Fig. 6a) and optimistic scenarios (Fig. 6b). It is possible to observe the evolution of the CO_2 emissions balance throughout time and accordingly the scenarios. The volume of RDF produced in the additional demand scenario varies between 183 and 992% higher than that of the base scenario, while the variations in network cost vary from 409 to 3399%. Lower fuel replacement implies less RDF produced compared to the total waste available in the base scenario, while the additional demand scenario maximizes the use of waste. For CO_2 equivalent emissions, the major discrepancies are not related to lower fuel replacement. In this case, it represents the balance of emissions from the reverse logistics network and the avoided emissions based on the different amounts of RDF produced in each scenario. For CO_2 equivalent emissions, the difference between the scenarios varies from 84 to 222%.

Table 4 Emissions of CO₂ produced and avoided

			Fuel substitution	Emissions (tonnes CO₂)				
				Reverse logistics network	Avoided in disposal	Avoided in fossil fuel supply	Avoided in fossil fuel replacement	Balance
Intermediate	2024	Base	15% CE	909.59	−618.02	−2440.28	−68.34	−2217.04
			30% CE	784.40	−618.02	−4880.55	−68.34	−4782.51
			50% CE	764.91	−618.02	−8134.26	−68.34	−8055.70
		Additional demand	15% CE	NA[a]	NA[a]	NA[a]	NA[a]	NA[a]
			30% CE	NA[a]	NA[a]	NA[a]	NA[a]	NA[a]
			50% CE	NA[a]	NA[a]	NA[a]	NA[a]	NA[a]
	2032	Base	15% CE	1066.24	−981.03	−2440.28	−92.08	−2447.14
			30% CE	2001.52	−2061.95	−4880.55	−184.17	−5125.15
			50% CE	4283.95	−3061.34	−8134.26	−306.94	−7218.58
		Additional demand	15% CE	2622.15	−3101.85	−2440.28	−297.69	−3217.67
			30% CE	2178.12	−3101.85	−4880.55	−297.68	−6101.97
			50% CE	NA[a]	NA[a]	NA[a]	NA[a]	NA[a]
	2040	Base	15% CE	1106.38	−1272.84	−2440.28	−92.08	−2698.82
			30% CE	2037.19	−2261.12	−4880.55	−184.17	−5288.65
			50% CE	3036.25	−3062.69	−8134.26	−306.94	−8467.64
		Additional demand	15% CE	2703.94	−4630.28	−2440.28	−440.38	−4807.00
			30% CE	2865.78	−4623.61	−4880.55	−440.38	−7078.76
			50% CE	3996.03	−4623.61	−8134.26	−440.38	−9202.21
Optimistic	2024	Base	15% CE	1029.55	−528.83	−2440.28	−92.08	−2031.64
			30% CE	1882.23	−1169.57	−4880.55	−168.50	−4336.39

(continued)

Table 4 (continued)

		Fuel substitution	Emissions (tonnes CO_2)				
			Reverse logistics network	Avoided in disposal	Avoided in fossil fuel supply	Avoided in fossil fuel replacement	Balance
	Additional demand	50% CE	2859.38	−1169.57	−8134.26	−168.50	−6612.94
		15% CE	2077.91	−1169.57	−2440.28	−168.50	−1700.43
		30% CE	NA[a]	NA[a]	NA[a]	NA[a]	NA[a]
		50% CE	NA[a]	NA[a]	NA[a]	NA[a]	NA[a]
2032	Base	15% CE	1083.15	−789.55	−2440.28	−92.08	−2238.76
		30% CE	1988.08	−2467.12	−4880.55	−168.50	−5528.09
		50% CE	2827.48	−3187.14	−8134.26	−168.50	−8662.41
	Additional demand	15% CE	4689.29	−6117.86	−2440.28	−602.72	−4471.57
		30% CE	4407.53	−6117.86	−4880.55	−602.72	−7193.61
		50% CE	4963.98	−6117.86	−8134.26	−602.72	−9890.87
2040	Base	15% CE	939.66	−948.86	−2440.28	−92.08	−2541.56
		30% CE	1920.84	−1128.88	−4880.55	−168.50	−4257.10
		50% CE	3357.59	−3867.32	−8134.26	−168.50	−8812.48
	Additional demand	15% CE	6094.07	−5620.89	−2440.28	−913.02	−2880.12
		30% CE	5570.08	−9215.65	−4880.55	−913.02	−9439.14
		50% CE	7054.23	−9215.65	−8134.26	−913.02	−11,208.69

[a] NA means not applicable

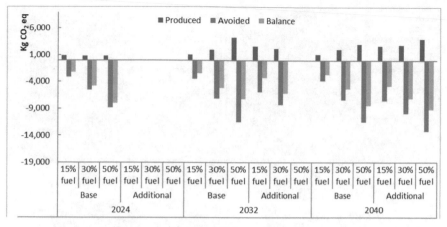

a) intermediate scenario

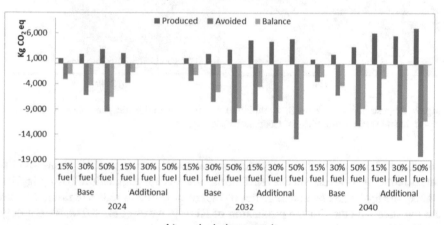

b) optimistic scenario

Fig. 6 Balance of GHG emissions in **a** intermediate and **b** optimistic scenarios

These results corroborate Silva et al. (2021). The authors analyzed the life-cycle assessment of RDF production from MSW in the city of Brasilia, the capital of Brazil. The production and use of RDF outperformed the conventional systems with emission reductions of 2% to about 23%. Avoided impacts by the replacement of petroleum coke by RDF in cement plants were more significant than recycling. However, this study considered the entire fraction of MSW, which includes the fraction of recyclables in RDF production. As this option contradicts the waste management hierarchy, in which recycling must be prioritized over WtE, this strategy will certainly face resistance from various sectors to be made viable.

In the initial years, the reverse logistics network operation is capillary to collect the necessary amount of RDF in a large portion of origin points to supply co-processing as evidenced in Fig. 7a. However, with the waste generation growth over time, it is

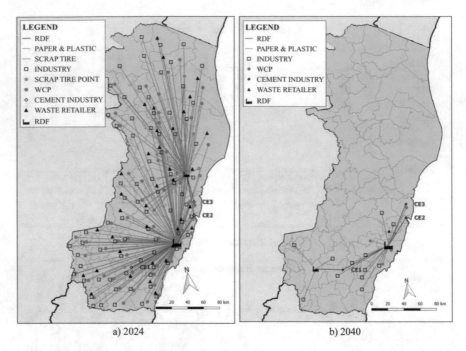

a) 2024   b) 2040

Fig. 7 Reverse logistics network evolution over time in optimistic scenario and 30% fuel replacement (WPC: Waste Pickers' Cooperative)

possible to concentrate the reverse logistics network closer to RDF plants significantly reducing travel distances even if the number of trips augments in 2040 (Fig. 7b). Hence, the optimized reverse logistics has great potential to reduce emissions from the logistics structure for the production of RDF (Chaves et al. 2021a, b), not only minimizing the environmental footprint but also reducing the costs of the proposal. This cost reduction for structuring RDF production is very important to enable its implementation or expansion in developing countries. Other benefits such as traffic reduction and the minimization of accidents on Brazilian highways would be other indirect advantages.

Cement production in Espírito Santo state achieved 753.85 million tonnes in 2018, which represents only 1.4% of the national industry (SNIC 2020). The Brazilian cement industry consists of 24 industrial groups that bring together 100 production units responsible for 53.46 million tonnes of cement in that year in 2018. Up to now, only the Votorantim group initiated the cement fuel replacement by RDF (Votorantim Cimentos 2019). This process was launched in 2019 representing the inaugural project in Brazil. Considering the development of the regulatory framework and the incentives that waste management plans point to its energy recovery, this case study demonstrates the potential for a minor share of the cement industry in the country. The impact could be extended to other regions, as Brazil is the world's 5th largest cement producer with a tendency to continue the sector expansion (CEMBU-REAU 2019). Gomes et al. (2019) indicate that the contribution of non-renewable fossil fuels in Brazilian cement production could reduce from 85 to 45% the fossil fuel participation, due to augmented use of waste and biomass in the fuel mix.

This potential is subject to uncertainties in the implementation of waste management policies as well as incentives for the development of the RDF industry. Under this speculative situation, corporations in the packaging sector, paper and plastics producers, and tire manufacturers can engage with the cement industry in sectoral agreements to promote the use of rejected waste for RDF production (Chaves et al. 2021a, b). This arrangement is beneficial for all parties involved, as the packaging and tire manufactures are concerned with waste materials by the extended producer responsibility (Hong and Guo 2019). However, Brazil is facing challenges to implement and assure compliance with this principle. The low adherence of retailers and importers, the considerable presence of free riders, fragile participation of municipalities, the absence of tax incentives, the low adhesion of generators, high operating costs, and collection problems are some barriers pointed by Ribeiro and Kruglianskas (2020) to effectively deploy extended producer responsibility in Brazil.

7 Final Remarks

The cement production footprint could be minimized with the RDF produced from rejected wastes. Co-processing with RDF from non-recyclable waste provides greater value to waste previously destined to landfills. In this sense, this proposal is aligned with the circular economy as it decreases the disposal of waste with energy potential in sanitary landfills and reduces GHG emissions. An industrial symbiosis via a mutually beneficial relationship between cement production and RDF production from residual waste can be established, providing economic benefits while minimizing their environmental impact (GHG emissions reduction). To ensure the sustainability of this proposal, companies of the packaging sector, tire manufacturers, and cement groups should establish partnerships sharing the necessary incentives with the government.

The development of the RDF industry in Brazil is benefiting from recent interest from government institutions. The uncertainties related to WtE and waste management policies implementation could affect the waste supply. A strong governance structure could favorise co-processing and RDF production processes. However, RDF facilities costs are significant and can compete with the low costs of waste landfilling and the petroleum coke price in Brazil. Since the production of RDF will be dependent on the cost–benefit of the project, future studies must address an economic assessment to evaluate its feasibility. considering the Brazilian reality. Only then would the environmental footprints of the cement industry be reduced with RDF use as alternative fuels.

The case study presented did not use more complete methodologies for environment footprint evaluation, such as Life-Cycle Analysis, for example. Further studies may improve the mechanisms for evaluating the environmental footprints of RDF production by considering a greater extent of impacts. It is worth considering the social footprint involved with this proposal. In Brazil, MSW screening is carried out mainly by waste pickers cooperatives. The waste involved in this study was only the rejected fraction from the sorting processes, which will potentially increase the gains of these cooperatives. This may result in more waste pickers formalized and/or increased gains from each of them. This is very important to consider in developing countries, where a significant portion of the population survives from the collection and sale of the recyclable fraction of MSW.

Even though describing the potential for Brazil, the benefits of RDF production both to minimize the environmental footprint of the cement industry but also to support rejected waste circular economy, can benefit other developing countries. However, it emphasizes the necessity to establish national standards for pollutant emissions control and governance to monitor compliance.

Acknowledgements The authors acknowledge the partial financial support of the Fulbright Scholar Program (award G-1-00005) from the Department of State, the United States of America, Federal University of Espírito Santo, and the National Council for Scientific and Technological Development—CNPq (Process 308411/2018-8 and 307835/2017-0) as well as the support from the University of Central Florida Global Program. Thanks, FICO for the free license of FICO® Xpress-IVE 8.8 software.

References

ABCP—Brazilian Association of Portland Cement (2019) Panorama of coprocessing 2017, São Paulo (in Portuguese). Available at: https://abcp.org.br/wp-content/uploads/2019/11/Panora maco_processamento_2019_v2-bx.pdf. Accessed Feb 2020

ABNT—Associação Brasileira de Normas Técnicas (2020) Resíduos sólidos urbanos para fins energéticos—Requisitos. Projeto ABNT NBR 16849. Available at: https://www.abntonline.com. br/consultanacional/projet.aspx?ID=28465. Accessed Jan 2020

ABRELPE—Brazilian Association of Urban Cleansing and Waste Management (2018) Panorama of solid waste in Brazil 2017. ABRELPE, Sao Paulo, Brazil (in Portuguese). Available at: http:// abrelpe.org.br/pdfs/panorama/panorama_abrelpe_2017.pdf

Al-Salem, SM (2019) Energy production from plastic solid waste (PSW). In: Al-Salem, SM (ed.). Plastics to energy. William Andrew Publishing, Elsevier, . pp 45–64

Ammenberg J, Baas L, Eklund M, Feiz R, Helgstrand A, Marshall R (2015) Improving the CO_2 performance of cement, part III: the relevance of industrial symbiosis and how to measure its impact. J Clean Prod 98:145–155

Bartholomeu DB, Péra TG, Caixeta-Filho JV (2016) Logística sustentável: avaliação de estratégias de redução das emissões de CO_2 no transporte rodoviário de cargas. J Transp Lit 10(3):15–19 (in Portuguese)

Bataille C (2020) Low and zero emissions in the steel and cement industries: barriers, technologies and policies. OECD green growth papers, No. 2020/02. OECD Publishing, Paris

Benhelal E, Zahedi G, Shamsaei E, Bahadori A (2013) Global strategies and potentials to curb CO_2 emissions in cement industry. J Clean Prod 51:142–161

Benhelal E, Shamsaei E, Rashid MI (2021) Challenges against CO_2 abatement strategies in cement industry: a review. J Environ Sci 104:84–101

Bourtsalas AT, Zhang J, Castaldi MJ, Themelis NJ (2018) Use of non-recycled plastics and paper as alternative fuel in cement production. J Clean Prod 181:8–16

Brazil (2019a) Law n° 274, April, 30, 2019. Disciplina a recuperação energética dos resíduos sólidos urbanos referida no § 1° do art. 9° da Lei n° 12.305, de 2010 (in Portuguese)

Brazil (2019b) Law n° 10.117, November, 19, 2019. Dispõe sobre a qualificação de projetos para ampliação da capacidade de recuperação energética de resíduos sólidos urbanos no âmbito do Programa de Parcerias de Investimentos da Presidência da República (in Portuguese)

Brazil (2020) Law n° 499, October, 6, 2020. Resolução CONAMA/MMA N° 499. Dispõe sobre o licenciamento da atividade de coprocessamento de resíduos em fornos rotativos de produção de clínquer (in Portuguese)

CEMBUREAU—The European Cement Association (2019) CEMBUREAU: activity report 2018— the cement sector: built in concrete made with cement. Available at: https://cembureau.eu/media/1818/actvity-report-2018.pdf

CETESB—Companhia Ambiental do Estado de São Paulo (2018) Estudo de baixo carbono para a indústria de cimento no estado de São Paulo de 2014 a 2030. CETESB, BID, São Paulo (in Portuguese). Available at: http://www.cetesb.sp.gov.br

Chandrasekhar K, Pandey S (2020) Co-processing of RDF in cement plants. In: Energy recovery processes from wastes. Springer, Singapore, pp 225–236

Chang YH, Chang NB (2001) Regional shipping strategy assessment based on installing a refuse-derived-fuel process in a municipal incinerator. Waste Manage Res 19(6):504–517

Chang NB, Pires A, Martinho G (2020) Solid waste management: life cycle assessment. In: Managing human and social systems. CRC Press, Boca Raton, pp 281–299

Chatterjee A, Sui T (2019) Alternative fuels—effects on clinker process and properties. Cem Concr Res 123:105777

Chaves GLD, dos Santos Jr JL, Rocha SMS (2014) The challenges for solid waste management in accordance with Agenda 21: a Brazilian case review. Waste Manag Res 32(9_suppl):19–31

Chaves GLD, Siman RR, Chang NB (2021a) Integrative policy analysis via system dynamic modeling for refuse-derived fuel production in Espírito Santo, Brazil. J Clean Prod (294):126344

Chaves GLD, Siman RR, Ribeiro GM, Chang NB (2021b) Synergizing environmental, social, and economic sustainability factors for refuse derived fuel use in cement industry: a case study in Espirito Santo, Brazil. J Environ Manag 288:112401

Dalmo FC, Simão NM, de Lima HQ, Jimenez ACM, Nebra S, Martins G, Palacios-Bereche R, de Mello Sant'Ana PH (2019) Energy recovery overview of municipal solid waste in São Paulo State, Brazil. J Clean Prod 212:461–474

DER-ES Departamento de Estradas de Rodagem do Espírito Santo (2019) Tabelas de distâncias (in Portuguese). Available at: https://der.es.gov.br/Media/der/Documentos/Rodovias%20Estaduais/TabelaDistancias.pdf

El-Salamony AHR, Mahmoud HM, Shehata N (2020) Enhancing the efficiency of a cement plant kiln using modified alternative fuel. Environ Nanotechnol Monit Manag 14:100310

EPE—Energy Research Company (2014) Energy consumption in Brazil: sectoral analysis. Empresa de Pesquisa Energética, Rio de Janeiro (in Portuguese). Available at: http://www.epe.gov.br/sites-pt/publicacoes-dados-abertos/publicacoes/PublicacoesArquivos/publicacao-251/topico-311/DEA2010-1420Consumo20de20Energia20no20Brasil[1].pdf

EPE—Energy Research Company (2018) Brazilian energy balance 2018 year 2017/Empresa de Pesquisa Energética. Empresa de Pesquisa Energética, Rio de Janeiro (in Portuguese). Available at: http://www.epe.gov.br/sites-pt/publicacoes-dados-abertos/publicacoes/PublicacoesArquivos/publicacao-303/topico-419/BEN2018__Int.pdf

EPE—Energy Research Company (2019) Brazilian energy balance 2019 year 2018/Empresa de Pesquisa Energética. Empresa de Pesquisa Energética, Rio de Janeiro (in Portuguese). Available at: http://www.epe.gov.br/pt/publicacoes-dados-abertos/publicacoes/balanco-energetico-nacional-2019

EPE—Energy Research Company (2020) Brazilian energy balance 2020 year 2019/Empresa de Pesquisa Energética. Empresa de Pesquisa Energética, Rio de Janeiro (in Portuguese).

Available at: https://www.epe.gov.br/sites-pt/publicacoes-dados-abertos/publicacoes/Publicaco
  esArquivos/publicacao-479/topico-528/BEN2020_sp.pdf
Ferdan T, Pavlas M, Nevrlý V, Somplâk R, Stehlík P (2018) Greenhouse gas emissions from thermal
  treatment of non-recyclable municipal waste. Front Chem Sci Eng 12:815–831. https://doi.org/
  10.1007/s11705-018-1761-4
Ferri GL, Chaves GDLD, Ribeiro GM (2015) Reverse logistics network for municipal solid waste
  management: the inclusion of waste pickers as a Brazilian legal requirement. Waste Manage
  40:173–191
Gallardo A, Carlos M, Bovea MD, Colomer FJ, Albarrán F (2014) Analysis of refuse-derived fuel
  from the municipal solid waste reject fraction and its compliance with quality standards. J Clean
  Prod 83:118–125
Galvez-Martos JL, Schoenberger H (2014) An analysis of the use of life cycle assessment for waste
  co-incineration in cement kilns. Resour Conserv Recycl 86:118–131
Garcés D, Díaz E, Sastre H, Ordóñez S, González-LaFuente JM (2016) Evaluation of the potential of
  different high calorific waste fractions for the preparation of solid recovered fuels. Waste Manage
  47:164–173
Genon G, Brizio E (2008) Perspectives and limits for cement kilns as a destination for RDF. Waste
  Manage 28(11):2375–2385
Georgiopoulou M, Lyberatos G (2018) Life cycle assessment of the use of alternative fuels in cement
  kilns: a case study. J Environ Manage 216:224–234
Gomes V, Cunha MP, Saade MRM, Guimarães GD, Zucarato L, Ribeiro CH, da Silva MG (2019)
  Consequential life cycle assessment of Brazilian cement industry technology projections for 2050.
  In: IOP conference series—earth and environmental science, vol 323, no 1. IOP Publishing, Bristol
Habert G, Miller SA, John VM, Provis JL, Favier A, Horvath A, Scrivener KL (2020) Environmental
  impacts and decarbonization strategies in the cement and concrete industries. Nat Rev Earth
  Environ 1(11):559–573
Haraguchi M, Siddiqi A, Narayanamurti V (2019) Stochastic cost-benefit analysis of urban waste-
  to-energy systems. J Clean Prod 224:751–765
Hashem FS, Razek TA, Mashout HA (2019) Rubber and plastic wastes as alternative refused fuel
  in cement industry. Constr Build Mater 212:275–282
Hoekstra AY, Wiedmann TO (2014) Humanity's unsustainable environmental footprint. Science
  344(6188):1114–1117
Hong Z, Guo X (2019) Green product supply chain contracts considering environmental responsi-
  bilities. Omega 83:155–166
Huang Q, Yang Y, Wang Q (2012) Potential for serious environmental threats from uncontrolled
  co-processing of wastes in cement kilns. Environ Sci Technol 46(24):13031–13032
IEA—International Energy Agency (2018) Technology roadmap—low-carbon transition in the
  cement industry. IEA. Available at: https://webstore.iea.org/technology-roadmap-low-carbon-tra
  nsition-in-the-cement-industry-foldout
IEA—International Energy Agency (2020) Cement. IEA, Paris. https://www.iea.org/reports/cement
IJgosse J (2019) Incineração de resíduos e Catadores: Um guia técnico sobre tecnologias de recu-
  peração energética de resíduos sólidos. Nota Técnica da WIEGO N°. 11. WIEGO, Manchester,
  UK
Infiesta LR, Ferreira CR, Trovó AG, Borges VL, Carvalho SR (2019) Design of an industrial solid
  waste processing line to produce refuse-derived fuel. J Environ Manage 236:715–719
IPCC (2014) Climate change 2014: mitigation of climate change. In: Edenhofer O, Pichs-Madruga
  R, Sokona Y, Farahani E, Kadner S, Seyboth K, Adler A, Baum I, Brunner S, Eickemeier P,
  Kriemann B, Savolainen J, Schlömer S, von Stechow C, Zwickel T, Minx JC (eds) Contribution
  of Working Group III to the fifth assessment report of the Intergovernmental Panel on Climate
  Change. Cambridge University Press, Cambridge, United Kingdom and New York, NY, USA
Jokar Z, Mokhtar A (2018) Policy making in the cement industry for CO_2 mitigation on the pathway
  of sustainable development—a system dynamics approach. J Clean Prod 201:142–155

Kajaste R, Hurme M (2016) Cement industry greenhouse gas emissions—management options and abatement cost. J Clean Prod 112:4041–4052

Karagulian F, Belis CA, Dora CFC, Prüss-Ustün AM, Bonjour S, Adair-Rohani H, Amann M (2015) Contributions to cities' ambient particulate matter (PM): a systematic review of local source contributions at global level. Atmos Environ 120:475–483

Kaza S, Yao LC, Bhada-Tata, P, Van Woerden F (2018) What a waste 2.0: A global snapshot of solid waste management to 2050. Urban development;. Washington, DC: World Bank

Kumar A, Dash SK, Ahamed MS, Lingfa P (2020) Study on conversion techniques of alternative fuels from waste plastics. In: Energy recovery processes from wastes. Springer, Singapore, pp 213–224

Lamas WQ, Palau JCF, de Camargo JR (2013) Waste materials co-processing in cement industry: ecological efficiency of waste reuse. Renew Sustain Energy Rev 19:200–207

Leal Filho W, Brandli L, Moora H, Kruopienė J, Stenmarck Å (2016) Benchmarking approaches and methods in the field of urban waste management. J Clean Prod 112:4377–4386

Lemc MMV, Rocha MH, Lora EES, Venturini OI, Lopes BM, Ferreira CH (2014) Techno-economic analysis and environmental impact assessment of energy recovery from Municipal Solid Waste (MSW) in Brazil. Resour Conserv Recycl 87:8–20. https://doi.org/10.1016/j.resconrec.2014.03.003

Lima PDM, Colvero DA, Gomes AP, Wenzel H, Schalch V, Cimpan C (2018) Environmental assessment of existing and alternative options for management of municipal solid waste in Brazil. Waste Manage 78:857–870

Liu X, Fan Y, Wang C (2017) An estimation of the effect of carbon pricing for CO_2 mitigation in China's cement industry. Appl Energy 185:671–686

Matuštík J, Kočí V (2020) What is a footprint? A conceptual analysis of environmental footprint indicators. J Clean Prod 124833

Miller SA, Moore FC (2020) Climate and health damages from global concrete production. Nat Clim Change 10(5):439–443

Miller SA, John VM, Pacca SA, Horvath A (2018) Carbon dioxide reduction potential in the global cement industry by 2050. Cem Concr Res 114:115–124

Mokrzycki E, Uliasz-Bocheńczyk A (2003) Alternative fuels for the cement industry. Appl Energy 74(1–2):95–100

Nejati V, Roknizadeh J (2014) Feasibility and economic evaluation of applying two alternative energy (RDF and TDF) in cement industries in Iran. In: Energy sustainability, vol 45875. American Society of Mechanical Engineers, p V002T04A004

Nevrlý V, Šomplák R, Putna O, Pavlas M (2019) Location of mixed municipal waste treatment facilities: cost of reducing greenhouse gas emissions. J Clean Prod 239:118003

Paolo M, Paola M (2015) RDF: from waste to resource–the Italian case. Energy Procedia 81:569–584

Papageorgiou A, Karagiannidis A, Barton JR, Kalogirou E (2009) Municipal solid waste management scenarios for Attica and their greenhouse gas emission impact. Waste Manage Res 27(9):928–937

Papanikola K, Papadopoulou K, Tsiliyannis C, Fotinopoulou I, Katsiampoulas A, Chalarakis E, Lytras GM (2019) Food residue biomass product as an alternative fuel for the cement industry. Environ Sci Pollut Res 1–10

PBMC—Painel Brasileiro de Mudanças Climáticas (2013) Relatório Executivo (in Portuguese). Disponível em: http://www.pbmc.coppe.ufrj.br. acessado em: 05/02/2016

Pires A, Martinho G (2019) Waste hierarchy index for circular economy in waste management. Waste Manage 95:298–305

Rada EC, Andreottola G (2012) RDF/SRF: which perspective for its future in the EU. Waste Manage 6(32):1059–1060

Rahman A, Rasul MG, Khan MMK, Sharma S (2015) Recent development on the uses of alternative fuels in cement manufacturing process. Fuel 145:84–99

Reza B, Soltani A, Ruparathna R, Sadiq R, Hewage K (2013) Environmental and economic aspects of production and utilization of RDF as alternative fuel in cement plants: a case study of Metro Vancouver Waste Management. Resour Conserv Recycl 81:105–114

Ribeiro FR, Kruglianskas I (2020) Critical factors for environmental regulation change management: evidences from an extended producer responsibility case study. J Clean Prod 246:119013

Rootzén J, Johnsson F (2017) Managing the costs of CO_2 abatement in the cement industry. Clim Policy 17(6):781–800

Samolada MC, Zabaniotou AA (2014) Energetic valorization of SRF in dedicated plants and cement kilns and guidelines for application in Greece and Cyprus. Resour Conserv Recycl 83:34–43

Sarc R, Lorber KE (2013) Production, quality and quality assurance of Refuse Derived Fuels (RDFs). Waste Manage 33(9):1825–1834

Scrivener KL, John VM, Gartner EM (2018) Eco-efficient cements: potential economically viable solutions for a low-CO_2 cement-based materials industry. Cem Concr Res 114:2–26

Silva V, Contreras F, Bortoleto AP (2021) Life-cycle assessment of municipal solid waste management options: a case study of refuse derived fuel production in the city of Brasilia, Brazil. J Clean Prod 279:123696

SIMA—Secretaria de Estado de Infraestrutura e Meio Ambiente (2019) Resolução SIMA N° 12, de 22 de fevereiro de 2019. Institui o Comitê de Integração de Resíduos Sólidos, e dá outras providências, São Paulo (in Portuguese)

SNIC—National Union of Cement Industries (2020) Produção nacional de cimento por regioes e estados em 2018. SNIC, Rio de Janeiro (in Portuguese). Acessado em outubro de 2019 em http://snic.org.br/assets/pdf/numeros/1565120104.pdf

Souza MM, Da Silva DS, Rochadelli R, Dos Santos RC (2012) Calorific power estimate and characterization of residues from harvesting and processing of Pinus taeda for energy purposes. Rev Floresta 42:325–334 (in Portuguese). http://doi.org/10.5380/rf.v42i2.26593

Stafford FN, Raupp-Pereira F, Labrincha JA, Hotza D (2016) Life cycle assessment of the production of cement: a Brazilian case study. J Clean Prod 137:1293–1299

Stripple H, Ljungkrantz C, Gustafsson T, Andersson R (2018) CO_2 uptake in cement-containing products: background and calculation models for IPCC implementation. Swedish Environmental Research Institute, Stockholm. Available at: https://www.ivl.se/download/18.72aeb1b0166c003cd0d64/1541160245484/B2309.pdf

Votorantim Cimentos (2019) Resíduos urbanos como fonte de energia para nossas fábricas (in Portuguese). Available at: https://www.votorantimcimentos.com.br/noticia/residuos-urbanos-como-fonte-de-energia-para-nossas-fabricas/

World Business Council for Sustainable Development (WBCSD)/International Energy Agency (IEA) (2009) Cement technology roadmap 2009—carbon emissions

Yang Z, Gao X, Hu W (2021) Modeling the air pollutant concentration near a cement plant co-processing wastes. R Soc Chem RSC Adv 11(17):10353–10363

Zhao L, Giannis A, Lam W-Y et al (2016) Characterization of Singapore RDF resources and analysis of their heating value. Sustain Environ Res 26:51–54

Ecological Footprint of Multi-silicon Photovoltaic Module Recycling

Dilawar Husain, Kirti Tewari, Manish Sharma, Akbar Ahmad, and Ravi Prakash

Abstract Solar photovoltaic (PV) modules are increasingly being adopted as a sustainable energy option to meet the growing demand for clean electricity. With declining costs, exponential growth in solar PV production and installation has been observed in the last two decades. However, this is expected to produce a huge quantity of scrap at the end of its operational life. Hence, a sharp rise in the need for recycling such scrap will emerge by 2030. The recycling of the End-of-Life (EoL) PV modules can be justified on the basis of its environmental impact and the costs involved. Recycling techniques assume to reduce the environmental impact of EoL phase of solar PV modules. Therefore, there is a need to quantitatively assess the Ecological Footprint (EF) of solar PV recycling. The present study estimates the EF of multi-crystalline silicon (multi-Si) photovoltaic modules recycling. The EF of recycling of 1 MW_p PV modules is estimated as 9.45 global hectare (gha). With an increase in production and installation of solar PV modules, the recycling impact will also be scaled accordingly.

Keywords Ecological footprint · Sustainability · Renewable system · Recycling · Waste management

D. Husain
Department of Mechanical Engineering, Maulana Mukhtar Ahmad Nadvi Technical Campus, Malegaon, Maharashtra, India

K. Tewari
Department of Mechanical Engineering, National Institute of Technology Sikkim, Ravangla, Sikkim, India

M. Sharma (✉)
Department of Mechanical Engineering, Malla Reddy Engineering College, Hyderabad, Telangana, India

A. Ahmad
Faculty of Science and Information Technology, Mianz International College, Male, Maldives

R. Prakash
Department of Mechanical Engineering, Motilal Nehru National Institute of Technology Allahabad, Prayagraj, Uttar Pradesh, India

1   Introduction

The problem of increasing power demand is also increasing with the electrification of rural places in India and abroad (Ahmad et al. 2017). Fossil fuel-generated electricity is responsible for 400–1000 g of CO_{2eq}/kWh but the emission from Si-based photovoltaic is negligible. India has steeply increased the installed capacity of solar power plants to 42 GW (Ahmad et al. 2015). The scenario is the same in other countries too for example Saudi Arabia approved a proposal of 300 MW PV plant which will produce power at a much cheaper rate of 0.0234 USD/kWh (Schmela 2018).

Since solar power is safe and reliable, thus there is an increase in capacity enhancement (Ahmad and Samuel 2016). The need of the hour is to have a sustainable energy source for power generation. Solar photovoltaic seems to be one of the sustainable sources. The policy of government to achieve self-sufficiency of power generation also revolves around the solar PV module (MNRE 2021). The other reason for this rapid change is the development of latest technology in the manufacturing of the PV module. The solar market is growing and along with it is growing the hazardous waste that will be generated at the end-of-life. This will also adversely affect the environment if they are not recovered and disposed properly (Shin et al. 2017). The EoL management of the solar panel is a growing issue of environmental sustainability. Thus, sustainable recycling is becoming more significant by the growing size of the installations around the world.

PV waste much ends up in landfills. A proper recycling process is required for the proper exertion of valuable resources. In order to maintain the sustainability of the PV power generation, low-cost recycling should synchronize with the rapid commercialization of these techniques. The current recycling only deals with the recovery of a part of the material or metals in the PV modules. The EF of 1 MW panels (Panel Capacity 320 W_p) production is estimated as 344.25 gha (Biswas et al. 2020). Another study reported that the average life cycle ecological footprint per m^2 collector area of the solar PV system has been evaluated as 0.0694 gha/m^2 (Husain et al. 2021). Hence there is still a lot of aspects of recycling that can be analysed.

Various countries are forming regulatory authorities for the recycling of PV panels and modules. Currently, only European Union have a strict and monitoring framework for recycling (Lunardi et al. 2018). The Waste Electrical and Electronic Equipment Directive 2012/19/EU was issued to protect and preserve the environment, human health while consuming the resources logically. In Japan, the government guarantee "feed-in Tariff" for electrical energy generated through renewable energy sources and thus there is a tremendous increase in solar installations. In order to regulate the recycling of the panels, the government has issued a voluntary guideline for proper disposal of the EOL panels. In the USA, resource preservation and recovery act regulate the disposal of harmful/hazardous materials. There is an amendment pending in the senate for assisting in the easy transport of the EOL panels. Most countries don't have the inclusion of PV waste in waste management policy due to different reasons. Since the life cycle of a PV module is assumed to be of 25 years and thus the waste size is still insignificant as compared to other waste management in these countries.

In present scenario, the landfills are inexpensive as compared to recycling methods of the PV modules. Thus, making recycling an uneconomical process and an unfavourable option for the silicon-based module as not many valuable materials are to be recovered (Sener and Fthenakis 2014). Recycling also has environmental impact, that can be analysed using the Ecological Footprint of the process.

2 Photovoltaic Technology

The term photovoltaic is often abbreviated as PV, P for Photo indicating light and V for Voltaic meaning electricity. Thus, PV is a means of converting the sunlight directly into electricity. This technology is considered to be easy to operate modular and highly reliable with maintenance-free installation. The principle of working is photoelectric effect, i.e. generation of electrical current when light falls on the photosensitive material connected to an electrical circuit. Figure 1 shows the basic schematics of PV cell construction. The top layer is encapsulated seal for protecting the solar cell. The cell consists of different layers of semiconductor materials having different electronic properties. These layers are created by doping of boron and phosphorous to give a positive and negative character. The transition of two doping is known as P/N junction as shown in Fig. 1. These individual cells are coupled to form a module and modules are connected in series and parallel to form an array. These arrays are solar thermal power plant generating units.

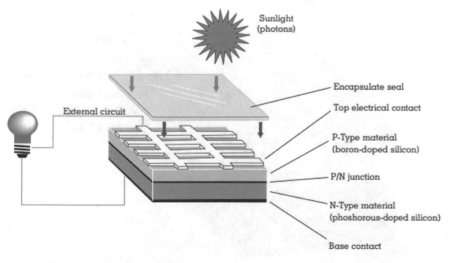

Fig. 1 Schematics of PV cell construction

2.1 Types of PV Modules

There are various types of PV modules available in the market and thus in order to understand the recycling process, the construction must be verified. The different PV modules are classified into

(a) Crystalline Silicon Module
(b) Thin Film Module
(c) Concentrating PV Module

(a) **Crystalline Silicon Module**: Crystalline Silicon (C-Si) Modules are market leaders with a market share of more than 90%. The aluminium back surface field (Narasinha and Rohatgi 1997) had been industrial norm for PV modules. Recently the passivated emitter and rear cell (Green 2015) has gained dominance in the world market and will replace the Al-BSF modules in the coming decade. The hetero-junction (HIT) cells are another type of cell gaining the interest of the manufacturing industries and forecast 15% of the total market share by 2027 (Weckend et al. 2016). Other cells are also making to market with mass production at low cost, including Si-based tandem solar cells (Weckend et al. 2016).

In a typical module, there are about 60–72 cells connected electronically in series to generate the required module. The components of C-Si Cells are shown in Fig. 2, weight distribution (Sander and Politik 2007; Wambach et al. 2005) is shown in Table 1. The recycling of materials was targeted to achieve 65% of recycling by weight but in some studies, it is shown that the recycling percentage can go up as high as 80% as in case of Lithuania and Bulgaria (Lunardi et al. 2018).

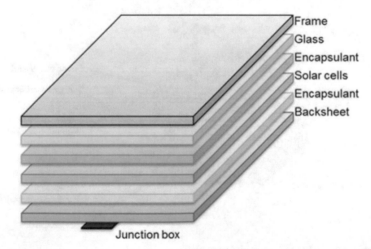

Fig. 2 Components of C-Si solar cell (Ranjan et al. 2011)

Table 1 Materials used in C-Si solar module

S. No.	Material	Component	Weight percentage
1	Glass	Module surface	75%
2	Polymer	Encapsulate and back sheet foil	10%
3	Aluminium	Frame	8%
4	Silicon	Solar cells	5%
5	Copper	Inter connectors	1%
6	Silver	Contact lines	0.1%
7	Tin and Lead	Miscellaneous components	Remaining

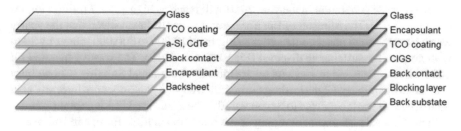

Fig. 3 Components of thin film solar cell (Ranjan et al. 2011)

(b) **Thin Film Module**: Thin Film Modules accounts for 10% of market share. There are three types of thin film cells in the market namely cadmium telluride (CdTe) with a share of 65%, copper indium gallium selenide (CIGS) with a share of 25% and amorphous silicon (a-Si) with a share of 10% in thin film market (FISE 2021).

In thin film cells, the cells are manufactured by depositing a thin layer of semiconductors (CdTe, CIGS, a-Si) on a substrate like glass, polymer or metal as shown in Fig. 3.

The idea for development of the thin film was to provide a low-cost alternative to C-Si solar cells along with added flexibility and lower material consumption. The most prominently used is CdTe with a major part of Cd (cadmium), which is toxic in nature and thus creates an environmental issue in disposal. The CIGS have higher efficiency due to higher bandgap. The recent studies conclude that the a-Si will be eliminated from the market due to its low efficiency and durability.

(c) **Concentrating PV Module**: CPV modules account for less than 1% of the market share and is a new technology, which enables the generation of electricity even at high temperatures. The research on the CPV is focused on generation of electricity with thermal power generation. The recycling CPV is yet to be started and thus the appropriate method is to be adopted in the coming years.

3   Global Market and Waste Generation

China is leading with 32% of installed solar power production. Followed by Japan generating 12% and India only 5%. The top 10 countries with an installed capacity of solar power generation are given in Fig. 4.

The market share of different solar panels and it expects to share in the future is projected and shown in Fig. 5. As indicated the market share of the C-Si is expected to fall and third-generation solar cell is expected to gain from this fall. The solar cell loses its efficiency over time in the life span of 25–30 years.

International Renewable Energy Agency estimated that the waste generated from the EOL solar panel was 0.25 million Metric tonnes globally (IRENA 2016). The expected waste generation in the year 2050 will reach as high as 9.57 million tonnes (Xu et al. 2018). The Minister of Environmental issues of Japan stated that the yearly waste from the solar panel generated is rising from 10,000 to 800,000 tonnes by 2040 with no plan for sustainable disposal (Fiandra et al. 2019). The waste generated till 2020 will be processed in 19 years by Toshiba Environmental Solution (Chowdhury et al. 2020). Similarly, China leading the installed capacity also does not have any plan for disposal of waste expected to be generated at the EOL. The USA is another country with no disposal plan except dumping of the waste in landfills. European Union also has only landfill plans which lead to degradation of soil and water contamination. This lack of policy is due to high lifetime of the solar panel but it can be calculated by $WY = IY + 25$, where WY and IY are waste generated year and installation year.

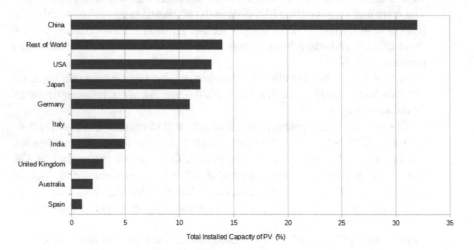

Fig. 4   Total installed capacity of solar power generation by end of year 2017 (IRENA 2016)

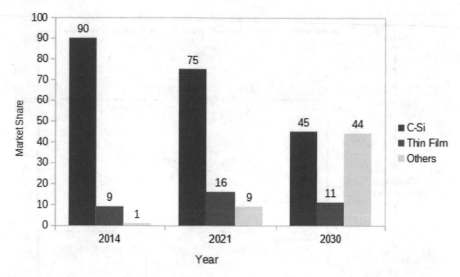

Fig. 5 Market share trend of different solar cell modules (Chowdhury et al. 2020)

4   Methods of Recycling PV Modules

Recently only two methods of recycling methods been in practice but other methods are also in the research stage and will soon be available for commercial use (Chowdhury et al. 2020). The process of recycling is different for different types of modules, i.e. C-Si is different from CdTe and CIGS. The Solar Modules are majorly recycled. The component materials like aluminium, copper, glass and semiconductor can be recovered and reused effectively. The recycling of modules progresses in three stages as shown in Fig. 6.

The process starts with removal of junction box cables and frames. The waste at the EOL is first separated by physical separation, thermal treatment and chemical treatment. The components like glass, aluminium and other components are collected and reused for new panel generation. The second stage of recycling is solar cells recovery to get the rare earth material like Silicon from C-Si and CIGS, CdTe and CIS from thin film solar cells.

A different organisation is working on different methods developed for recycling the PV module. PV Cycle is a non-profit group that was first to develop its own method of waste management for waste from PV modules. They manage to achieve a rate of 96% recycling of solid waste fraction (Cycle 2020). First solar developed a recycling process that is more suitable for the thin film modules.

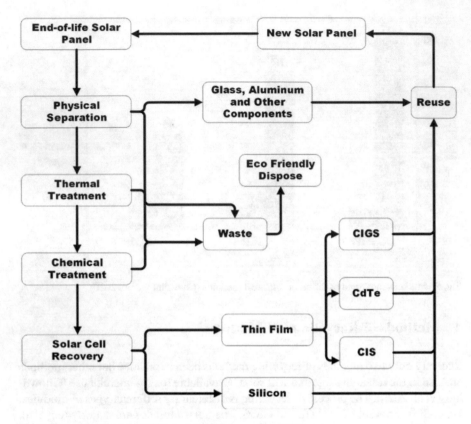

Fig. 6 Recycling process of PV modules (Waseem et al. 2020)

5 Ecological Footprint (EF)

Ecological Footprint (EF) indicator is one of the prominent tools that measure human activity (i.e. direct and indirect) in conjunction with regenerative capacity of the planet. The general expression for evaluating EF is as follows (Wackernagel and Rees 1996):

$$EF = \sum \left(\frac{C_i}{Y_i} \right) . e_i \tag{1}$$

where C_i is annual consumption of ith product (tonnes per year), Y_i is yield factor (annual productivity) of ith product (tonnes per ha), The e_i represents the equivalence factor (gha/ha) of different types of bio-productive lands, it mentioned in Table 2.

Carbon sequestration such as emission during material production, fossil fuel burning and transportation, etc. can be expressed as follows (GFN 2010):

Table 2 Equivalent factor of different types of bio-productive lands

Parameters		Value	References
Equivalence factor (e_i)	Cropland	2.52 gha	GFN (2016)
	Pastureland	0.43 gha	
	Forestland/CO_2 land	1.28 gha	
	Marine land	0.35 gha	
A_{oc}		0.3	SIO (2017)
A_f		2.68 tCO_2/ha	Husain and Prakash (2019)

$$\text{EF of carbon sequestration} = X_c \frac{(1 - A_{oc})}{A_f} . e_i \qquad (2)$$

where X_c is annual CO_2 emissions (tCO_2), A_{oc} is the fraction of annual oceanic anthropogenic CO_2 sequestration and A_f is the annual rate of carbon uptake/hectare of forestland at world average yield.

Limitations

The study has some limitations that are required for Ecological Footprint Assessment of multi-Si module recycling. The details are mentioned as follows:

(a) literature data are used to assess module recycling
(b) materials transportation is not considered
(c) assumptions are used for feasibility assessment, etc.

Objective

The objective of this study is to evaluate the ecological impact of multi-Si module recycling processes. The EF analysis has been used to assess the recycling impact of module. The study will help in the overall estimation of resource consumption and environmental impact after the EoL of multi-Si modules. Such a study may also be helpful in exploring the feasibility as well as reduction potential in EF recycling of solar photovoltaic power generation systems in the country.

6 Methodology

The Ecological Footprint assessment of solar PV panels recycling and their quantitative environmental impact (after the end-of-life phase) on the planet may give clear and comprehensive data to encourage appropriate product recycling policies. The solar PV panel waste flow that wants to be reused/recycled/disposed can be estimated and the composition of the panel waste can be calculated. The crystalline silicon-based module recycling is estimated in this case study, which dominates the

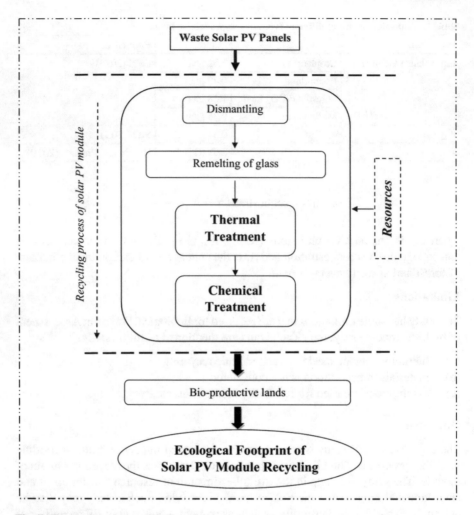

Fig. 7 System boundaries of C-Si modules

solar power production market (approximate 85% of the total power generation). The system boundaries of this research have shown in Fig. 7.

6.1 Ecological Footprint of Solar PV Panel Recycling (EF_R)

The study illustrates the Ecological Footprint of silicon-based solar PV panel recycling. Solar PV panel recycling has a significant environmental impact, therefore, to assess the environmental impact of panel recycling needs to calculate the impact of all processes involved in recycling. In general, the four processes are involved in panel recycling: (1) Dismantling, (2) Remelting of glass, (3) Thermal treatment and (4) Chemical treatment. The silicon-based panel recycling depends on these four parameters. The EF of solar PV panel recycling is as calculated in Eq. (3).

$$EF_R = EF_{dismantling} + EF_{remelting} + EF_{thermal} + EF_{chemical} \tag{3}$$

where, $EF_{dismantaling}$ represent the Ecological Footprint of dismantling process, $EF_{remelting}$ represent the Ecological Footprint of remelting of glass, $EF_{thermal}$ represents the Ecological Footprint of thermal treatment process, $EF_{chemical}$ represents the Ecological Footprint of chemical treatment process.

6.1.1 Ecological Footprint of Dismantling ($EF_{dismantling}$)

Solar PV module recycling initiates with the manual dismantling process that helps to categories the waste materials for recycling. However, the procedure is unable to fragment different waste materials in an efficient manner. The Ecological Footprint of dismantling is estimated as Eq. (4)

$$EF_{dismantling} = \sum W_i D_i . \alpha_i . \frac{(1 - A_{oc})}{A_f} . e_i \tag{4}$$

where, W_i represents the amount of solar PV module waste (i.e. unit of kg, m^2 or kW), D_i represents the energy required for the dismantling of the solar PV modules (MJ per unit of modules), α_i represents the emission factor of corresponding energy use (tCO$_2$/MJ).

6.1.2 Ecological Footprint of Remelting of Glass ($EF_{remelting}$)

The contribution of glass in a typical silicon PV module is about 75% of the total weight of the panel (PV Magazine 2020). The Solar PV panel recycling includes remelting of glass that is also accountable for ecological burdens on the planet, however, the ecological impact of remelting process is comparatively lower than the impact of landfill treatment of waste glass. The Ecological Footprint of remelting of glass is estimated as Eq. (5).

$$EF_{remelting} = \sum W_i R_i . \beta_i . \frac{(1 - A_{oc})}{A_f} . e_i \tag{5}$$

where, R_i represents the energy requirement for the remelting of glass of waste modules (MJ per unit of modules), β_i represent the emission factors of corresponding energy use in remelting glass of waste modules (tCO$_2$/MJ).

6.1.3 Ecological Footprint of Thermal Treatment (EF$_{thermal}$)

The recycling of crystalline Si PV module includes thermal treatment process that emits 370 kg CO_{2eq}/ton of PV waste or primary energy consume 2780 MJ/ton of PV waste (Latunussa et al. 2016). The thermal treatment is responsible for GHG emissions and primary energy use; however, the overall environmental impact is significantly less than the Si PV module production with virgin materials. The Ecological Footprint of thermal treatment is estimated as Eq. (6)

$$EF_{thermal} = \sum W_i.\mu_i + E_i\gamma_i.\frac{(1 - A_{oc})}{A_f}.e_i \tag{6}$$

where M_i represents the quantity of materials used during thermal treatment (kg/unit of module), μ_i represents the corresponding embodied energy of the material used during thermal treatment (tCO$_2$/kg of material), E_i represents the direct energy/fossil fuels consumption during thermal treatment (MJ/unit of module recycling; or kg of fossil fuel/unit of module recycling), γ_i represents the emission factor of direct energy/fossil fuel use (tCO$_2$/MJ; or tCO$_2$/kg of fossil fuel use).

6.1.4 Ecological Footprint of Chemical Treatment (EF$_{chemical}$)

The chemical treatment process is essential for Al (aluminium), Si and Ag (silver) before recycling in polysilicon production and PV cell, respectively. The chemical treatment process has some ecological impact but it reduces the impact of new module production. The Ecological Footprint of chemical treatment is estimated as Eq. (7).

$$EF_{chemical} = \sum Ni.\mu_j + C_j\gamma_j.\frac{(1 - A_{oc})}{A_f}.e_i \tag{7}$$

where N_j represents the quantity of materials used during chemical treatment (kg/unit of module), μ_j represents the corresponding embodied energy of the material used during chemical treatment (tCO$_2$/kg of material), C_j represents the direct energy/fossil fuels consumption during chemical treatment (MJ/unit of module recycling; or kg of fossil fuel/unit of module recycling), γ_j represents the corresponding emission factor of direct energy/fossil fuel use during chemical treatment (tCO$_2$/MJ; or tCO$_2$/kg of fossil fuel use).

7 Results

The environmental assessment of 1 MW$_P$ capacity multi-Si photovoltaic modules recycling has been performed in this study. The EF of modules recycling is estimated as 9.45 gha (i.e. 0.112 gha per tonne of module waste). The general recycling

Table 3 Calculation of $EF_{remelting}$

Recycling process of 1 MW (approx. 84 t) multi-Si modules	Input	Quantity (tonne)	Embodied emission (tCO$_2$/tonne)	Emission factor	EF (gha)
Dismantling	Multi-Si	84	5.5 kWh	0.82 tCO$_2$/MWh; MPCEA (2020)	0.18182
Remelting of glass (47.5 tonne)	Oxidant	9.47	0.41		1.28848
	Standard Coal	6.8	0.0045		0.010253
Thermal treatment (Ethylene vinyl acetate (15.4 tonne) produce during dismantling process)	N$_2$	16.2	0.43		2.31168
	Oxidant	9.63	0.41		1.31025
	Electricity		3 MWh	0.82 tCO$_2$/MWh; MPCEA (2020)	1.17
Chemical treatment for Al (including remelting phase), Ag and Si (approx. 15.23 tonne)	HNO$_3$	0.0223	1.55		0.01147
	NaOH	5.5	1.12		2.04421
	Standard coal	5.1	0.0045		0.00769
	HF	1.2	2.82		1.12299
Total					9.45

process of Si based photovoltaic modules is depicted in Table 3. It includes the Ecological Footprint of different processes of modules recycling. The details of the EF of recycling are as follows:

7.1 Ecological Footprint of Dismantling (EF$_{dismantling}$)

The energy demand for dismantling of solar modules is about 5.5 kWh/tonne (Ardente et al. 2019). The Ecological Footprint of dismantling of modules is about 0.18 gha per MW$_P$ capacity of the module, it contributes about 2% of the total recycling impact (Fig. 8). The details of the input parameters that use to calculate EF$_{dismantling}$ are mentioned in Table 3.

7.2 Ecological Footprint of Remelting of Glass (EF$_{remelting}$)

The Ecological Footprint of remelting of glass is estimated as 1.30 gha per MW$_P$ capacity of the module, it contributes about 14% of the total recycling impact (Fig. 8). The details of the input parameters that use to calculate EF$_{remelting}$ are mentioned in Table 3.

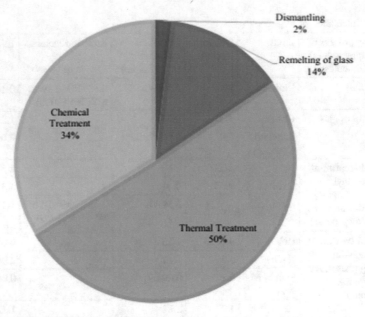

Fig. 8 Total recycling impact

7.3 Ecological Footprint of Thermal Treatment (EF_thermal)

The Ecological Footprint of thermal treatment process is calculated as 4.79 gha per MW_P capacity of the module, it contributes highest among all the recycling processes (i.e. half of the total recycling impact Fig. 8). The details of the input parameters that use to calculate $EF_{thermal}$ are mentioned in Table 3.

7.4 Ecological Footprint of Chemical Treatment (EF_chemical)

The Ecological Footprint of chemical treatment process is calculated as 3.20 gha per MW_P capacity of module, it contributes approximately one-third of the total recycling impact of the modules (Fig. 8). The details of the input parameters that use to calculate $EF_{chemical}$ are mentioned in Table 3.

In silicon module recycling, the thermal treatment contributes highest (approx. 50% of the total recycling impact) among all the processes followed by chemical treatment (i.e. 34%). The remaining two processes: remelting of glass and dismantling are contributed 14% and 2% of the total recycling impact, respectively. The results indicate that the environmental impact of multi-Si module recycling is about 2.7% of the module production.

According to the International Renewable Energy Agency (IRENA 2016), the global cumulative solar PV module waste was account for 0.1 million tonnes by the

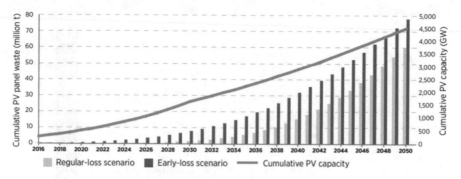

Fig. 9 Estimated cumulative global waste volumes (million t) of EOL PV modules (IRENA 2016)

end of 2020. It will increase significantly up to 1.7 million tonnes in 2030 (projected value), while a drastic module waste rise that could be approximately 60 million tonnes by 2050 (Fig. 9) A simulation model had been developed by IRENA to estimate the ratio of Solar PV module waste and new solar PV module installation.

The simulation results are depicted in Fig. 10. This ratio is negligible at the end of 2020 because annual solar PV module waste estimated at 0.1 million tonnes, while new solar PV module installation estimated at 5 million tonnes. However, it increases in future to 4% in 2030, 39% in 2040 and 89% in 2050. In Figure, approximately 6 million tonnes of solar PV module waste are forecast in 2050 with 7 million tonnes of new solar PV module installations.

According to IRENA report, the market share of C-Si module is about 73.3% in 2020. considering the same market share of C-Si Module till 2050 the projected ecological footprint of C-Si module recycling is depicted in Fig. 11.

In Fig. 11, the projected impact on ecological footprint of C-Si module waste recycling increased to as high as 6,772,460, i.e. approximately 600-fold of the impact in year 2020 in just three decades.

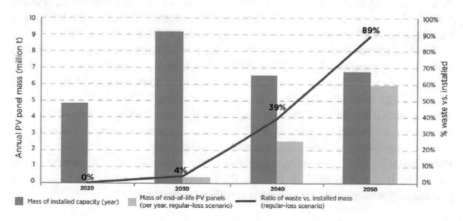

Fig. 10 Projection of module installation versus waste generation (IRENA 2016)

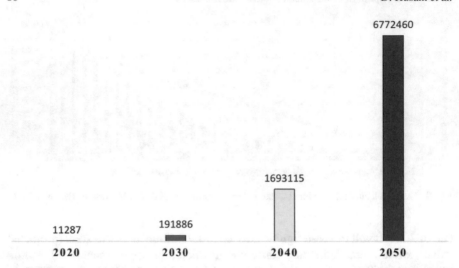

Fig. 11 Projected EF of C-Si module waste recycling

8 Conclusions

The case study contributes to advancing the information about the recycling processes of photovoltaic silicon-based modules, using literature data. The environmental impacts of solar PV modules recycling are reducing the overall silicon-based module production because the recovered materials (i.e. during recycling) can be reused for the new module's production. After the end-of-life of photovoltaic module, the general recycling process including: (1) dismantling, (2) remelting of glass, (3) thermal treatment and (4) chemical treatment; will result in a significant environmental imprint on the planet. The environmental assessment of 1 MW_P capacity multi-Si photovoltaic modules recycling has been performed in this study. The EF of modules recycling is estimated as 9.45 gha (i.e. 0.112 gha per tonne of module waste). The thermal treatment contributes highest (approx. 50% of the total recycling impact) among all the processes followed by chemical treatment (i.e. 34%).

Significant growth is predicted in coming decades, leading to an expected installed capacity of approximately 4.5 TW in 2050. Silicon-based modules have an average life expectancy of 25 years; most of the countries have started photovoltaic installation at a large scale after the end of the twentieth century. This study also forecasts that significant amounts of module waste will be generated by 2030 and 2050 as these solar PV systems recached their age. Based on the results, it indicates the possible improvement in recycling technology. It will help to support the research on recycling of multi-Si photovoltaic modules as the end-of-life of silicon modules in the world. It should be made to improve the recovery rate and motivate to use more eco-friendly materials in the different recycling processes.

References

Ahmad A, Samuel P (2016) Digital control of photovoltaic based multilevel converter for improved harmonic performance. Indian J Sci Technol 9(44). http://doi.org/10.17485/ijst/2016/v9i44/101957

Ahmad A, Samuel P, Amar Y (2015) Solarizing India: tapping the excellent potential. Renew Energy 9(3):13–17

Ahmad A, Khandelwal A, Samuel P (2017) Golden band search for rapid global peak detection under partial shading condition in photovoltaic system. Sol Energy 157:979–987

Ardente F, Latunussa CEL, Blengini GA (2019) Resource efficient recovery of critical and precious metals from waste silicon PV panel recycling. Waste Manage 91:156–167. https://doi.org/10.1016/j.wasman.2019.04.059

Biswas A, Husain D, Prakash R (2020) Life-cycle ecological footprint assessment of grid-connected rooftop solar PV system. Int J Sustain Eng. http://doi.org/10.1080/19397038.2020.1783719

Chowdhury MS, Rahman KS, Chowdhury T, Nuthammachot N, Techato K, Akhtaruzzaman M, Tiong SK, Sopian K, Amin N (2020) An overview of solar photovoltaic panels' end-of-life material recycling. Energ Strategy Rev 27:100431

Cycle P (2020) Annual report. https://pvcycle.org/wp-content/uploads/2021/05/2016-Annual-Report-PV-CYCLE-AISBL.pdf. Accessed on 05-06-2021

Fiandra V, Sannino L, Andreozzi C, Graditi G (2019) End-of-life of silicon PV panels: a sustainable materials recovery process. Waste Manag 84:91–101

Fraunhofer Institute for Solar Energy Systems (FISE). Photovoltaics report. www.ise.fraunhofer.de. Accessed on10 June 2021

Global Footprint Network (GFN) (2010) Calculation methodology for the national footprint accounts, 2010 edn. http://www.footprintnetwork.org/content/images/uploads/National_Footprint_Accounts_Method_Paper_2010.pdf. Accessed 21 June 2021

Global Footprint Network, GFN (2016) http://data.footprintnetwork.org/analyzeTrends.html?Cn=100&type=EFCtot. Accessed June 2021

Green MA (2015) The passivated emitter and rear cell (PERC): from conception to mass production. Sol Energy Mater Sol Cells 143:190–197

Husain D, Garg P, Prakash R (2021) Ecological footprint assessment and its reduction for industrial food products. Int J Sustain Eng 14(1):26–38. https://doi.org/10.1080/19397038.2019.1665119

Husain D, Prakash R (2019) Ecological footprint reduction of building envelope in a tropical climate. J Inst Eng (India) Ser A 100:41–48. http://doi.org/10.1007/s40030-018-0333-4

International Renewable Energy Agency (IRENA) (2016) End-of-life-management—solar photovoltaic panel. ISBN 978-92-95111-98-1

Latunussa CEL, Ardente F, Blengini GA, Mancini L (2016) Life cycle assessment of an innovative recycling process for crystalline silicon photovoltaic panels. Sol Energy Mater Sol Cells 156:101–111

Lunardi MM, Alvarez-Gaitan JP, Bilbao JI, Corkish R (2018) A review of recycling processes for photovoltaic modules, solar panels and photovoltaic materials, pp 9–27

Ministry of New and Renewable Energy (MNRE), Government of India. Initiatives and achievements. https://mnre.gov.in/. Retrieved on 15th June 2021

Ministry of Power Central Electricity Authority, Government of India (MPCEA) (2020) CO_2 baseline database for the Indian power sector. User guide 2016. Accessed Nov 2020

Narasinha S, Rohatgi A (1997) Optimized aluminum back surface field techniques for silicon solar cells. In: Photovoltaic specialists conference, 1997. Conference record of the twenty sixth IEEE. IEEE, pp 63–69

PV Magazine (2020) Solar panel recycling: turning ticking time bombs into opportunities. https://www.pvmagazine.com/2020/05/27/solar-panel-recycling-turning-ticking-time-bombs-into-opportunities/

Ranjan S, Balaji S, Panella RA, Ydstie BE (2011) Silicon solar cell production. Comput Chem Eng 35(8):1439–1453

Sander K, Politik I (2007) Study on the development of a take back and recovery system for photovoltaic products

Schmela M (2018) Solar power Europe. Global market outlook for solar power: 2018–2022

Scripps Institution of Oceanography (SIO), The Keeling Curve (2017). https://scripps.ucsd.edu/programs/keelingcurve/2013/07/03/how-much-co2-can-the-oceans-take-up/. Accessed on April 2017

Sener C, Fthenakis V (2014) Energy policy and financing options to achieve solar energy grid penetration targets: accounting for external costs. Renew Sustain Energy Rev 32:854–868

Shin J, Park J, Park N (2017) A method to recycle silicon wafer from end-of-life photovoltaic module and solar panels by using recycled silicon wafers. Sol Energy Mater Sol Cells 162:1–6

Wackernagel M, Rees W (1996) Our ecological footprint: reducing human impact on the earth. New Society, Gabriola Island, British Columbia. ISBN:1-55092-251-3

Wambach K, Müller A, Alsema E (eds) (2005) Life cycle analysis of a solar module recycling process. In: 20th European photovoltaic solar energy conference, Barcelona, Spain

Waseem R, Brahambhatt JI, Patel R (2020) A review paper on growing photo voltaic cell waste, its handling, recovery and disposal. Solid State Technol 63(6)

Weckend S, Wade A, Heath G (2016) End-of-life management—solar photovoltaic panels. Report of the International Renewable Energy Agency (IRENA) and the International Energy Agency (IEA)

Xu Y, Li J, Tan Q, Peters AL, Yang C (2018) Global status of recycling waste solar panels: a review. Waste Manag 450–458

An Environmental Construction and Demolition Waste Management Model to Trigger Post-pandemic Economic Recovery Towards a Circular Economy: The Mexican and Spanish Cases

Pilar Mercader-Moyano, Jesús López-López, and Patricia Edith Camporeale

Abstract The Architecture, Engineering and Construction (AEC) industry consume 40% of raw material generating 35% of industrial waste worldwide. In the EU, it consumes 50% of raw material and generates 35% of industrial waste; while in the USA, 22% of 600 million tons of Construction and Demolition Waste (CDW) were recycled into new products: 52% as aggregates but 24% ended in landfills, in 2020. A transition to a circular economy may trigger post-COVID-19 economic recovery. Thus, the EU promotes it through the EU Green Deal, Renovation Wave and Circular Economy Action Plan. This work applies the Spanish CDW "weighted transfer of measurement" current model to broaden its construction material database and to add environmental indicators. Latin America has the world's highest urbanization rate (84%) but lacks effective CDW management to thrive in Regenerative Sustainability, Climate Change mitigation and post-pandemic economic recovery. This research quantifies onsite 61 Mexican social housing CDW, comparing both countries' results. Mexico consumes 1.24 ton m^{-2} of raw materials and produces 0.083 ton m^{-2} CDW with a 16% recycling rate, while Spain consumes 1.90 ton m^{-2} and produces 0.08 ton m^{-2} with a 75% recycling rate. Cement-based, ceramic and mixed CDW represent 83.44% for Mexico and 95.61% for Spain. The implementation of this methodology will deliver sustainable CDW management in Mexico, minimizing CDW production, by the replacement of current construction materials for eco-efficient ones and the promotion of related legislation. Moreover, this updated

P. Mercader-Moyano (✉)
Departamento de Construcciones Arquitectónicas I. Escuela Técnica Superior de Arquitectura, Universidad de Sevilla, Avenida Reina Mercedes 2, 41012 Seville, Spain
e-mail: pmm@us.es

J. López-López
Facultad de Arquitectura, Diseño y Urbanismo, Universidad Autónoma de Tamaulipas, Tampico, Mexico

P. E. Camporeale
Facultad de Arquitectura y Urbanismo, Universidad Nacional de La Plata, La Plata, Argentina

© The Author(s), under exclusive license to Springer Nature Singapore Pte Ltd. 2022
S. S. Muthu (ed.), *Environmental Footprints of Recycled Products*,
Environmental Footprints and Eco-design of Products and Processes,
https://doi.org/10.1007/978-981-16-8426-5_4

transformation coefficient database widens the Spanish model to an international scale.

Keywords Building sector circular economy · CDW environmental footprint assessment · CDW quantification · Post-pandemic economic recovery · CDW executive project indicators

1 Introduction

Buildings consume 40% of natural resources and primary energy worldwide (López-Mesa et al. 2009) while the AEC industry generates 35% of the industrial waste (Hendriks 2000) and 36% of Greenhouse Gas (GHG) Emissions. Only in the EU, the AEC industry represents 10% of GDP, consumes 50% of natural resources and 40% of primary energy and generate 35% of CDW. The carbon footprint accounts for 30% of the total but if construction materials had a more efficient use they could decrease 80% (Da Costa-Gómez 2020; Mercader Moyano 2010).

In the EU, CDW represents the largest flow in terms of mass: 1/3 of 3 billion tons annually, being this quantity relatively stable. This sector seems to be economically circular as it avoids landfills and incineration. However, CDW recycling mainly orients to backfilling with a low grade of recovery like recycled aggregates used in road subsoil, reducing its potential towards circular CWD management (European Commission 2018). Spain is the 7° EU country in the CDW production ranking: 138 million tons generated during 2018 with a 75% recycling rate (Eurostat 2020a) while the European average was 90% (Eurostat 2020b).

However, CDW is not suitable for reuse or recycling because of old construction techniques that prevent high purity raw material recovery during demolition or refurbishment. It could be possible to prevent CDW augment and achieve a better recycling process if prices became competitive, second-use materials were reliable and material data from existing buildings were available. Notwithstanding, the decades between the construction period and CDW management at the demolition or refurbishment stages constitute another barrier (European Environment Agency 2020).

COVID-19, declared a pandemic by the World Health Organization (WHO), provoked a sanitary crisis triggering the economy to the lowest activity level since 2009: GDP fell 1.674% worldwide (OECD 2020; World Bank 2020). Raw material and oil prices fall have become a barrier to these initiatives: idle assets, disruptions in supply chains, uncertainty in availability and price volatility add other barriers to the circular transition (Fig. 1).

However, the circular economy path to achieve decarbonized cities, energy poverty reduction and socioeconomic sustainability like the EU Circular Economy Action Plan (European Commission 2019), the "Renovation Wave" and the "Green Deal" (European Commission 2020a) show strong affinity to EU strategic priorities to economic recovery. These new business models activate local resources, reduce import dependence, diversify supplies to increase resilience and make create

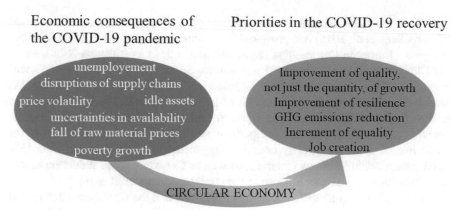

Fig. 1 The shift from the linear to the circular economy trigger post-pandemic recovery. *Source* The authors

650,000–700,000 jobs by 2030. The reduction of cost mobility and food benefits low-income households, promoting equality. In addition, circular economy contributes to achieving carbon neutrality with a nearly 300 million tons annual reduction (almost 50%) by 2050.

Furthermore, the EU Social and Economic Committee identifies the building sector as the main actor to promote European recovery after the COVID19 crisis because of the intensive use of labour in the AEC industry, mainly in the hands of local companies (Zahradnik et al. 2020). In Spain, the building energy retrofit plan manages a € 300 million budget provided by the National Fund for Energy Efficiency and EU Funds for the Spanish COVID economic recovery. That will generate around 48,000 jobs annually in the next nine years (Gobierno de España 2020); these funds will help the most vulnerable groups make the necessary housing retrofit including active or passive HVAC-SDHW systems. As a matter of fact, the circular economy promotes the 7 Rs: redesign, reduce, reuse, repair, renovate, recover and recycle (European Commission 2020b), making EU compel the standardization of second-use materials and its dissemination among stakeholders to reduce CDW production and acquire a better and higher recycling quality (European Commission 2020a). European Commission has issued 54 measures to apply along the construction material lifecycle in five prior sectors that comprehend construction and demolition. Some of them are the CDW Protocol and Guidelines (European Commission 2018) that improves the reliance on recycled products, Level(s) that provides an assessment and reporting framework to promote a lifecycle approach for residential and office buildings, one-use plastic ban, critical raw material reuse/recycling, eco-design, eco-labelling to favour spare parts availability for repair, waste treatment at the end of the lifecycle and packaging reuse/recycling (European Commission 2020c).

Nevertheless, after the 2008 crisis when Spanish regulations promoted building retrofit, they did not care about CDW plastic insulation material management. Consequently, in 2010, 10% of CDW (860 million tons) came from plastic materials apart from those typically associated, whose environmental footprint remains disregarded (Villoria Sáez et al. 2018) and provoked the failure to achieve the 70% CDW recycling rate (European Commission 2008; Ministerio de la Presidencia 2008). Since EU Directives guide EU members' regulations, Spain updated CDW management through the Spanish Strategy on Circular Economy 2030 (Jefatura de Gobierno 2011; Ministerio de Agricultura, Alimentación y Medio Ambiente 2015a, b; Ministerio de la Presidencia 2008). Furthermore, Spain intends to achieve a sustainable, decarbonized, resource-efficient and competitive economy consisting of six goals for 2030: five of them are directly or indirectly related to CDW management: 15% CDW reduction, 30% of raw materials, as well as CO_2e emissions reduction to less than 10 million annual tons because of the construction material reuse.

The situation is quite different In Latin America and the Caribbean; this region presents the highest urbanization rate worldwide (84%), where 32% of the total population lives in cities of more than 1 million people that accounts for 40% of the global urban population. Municipal solid waste (MSW) including CDW collection leaves aside 7% of marginal and rural people (40 million) that produce 35,000 tons daily (United Nations Environment Programme 2018). Moreover, 60% of MSW (145,000 tons daily) ends in landfills, some of which have acquired international standards as authorized places being most of them only "supervised landfills". In the case of Mexico City, which has got only two authorized landfills, the city government has launched the Zero Waste Plan to reduce the 8600 tons sent to landfills to 2000 tons by 2024, from a current amount of 16,000 daily tons (CDMX 2019; Ríos 2019). In Mexico, CDW accounts for 6.7% of GDP, generated approximately 5.6 million jobs (Cámara Mexicana de la Industria de la Construcción 2016) without including CDW due to natural disasters (earthquakes) (Araiza-Aguilar et al. 2019). Mexico, which is the second regional economy and the 15th globally, had a poverty rate of 48.8% in 2018 (61.1 million people) that will increase to 66.9%, becoming the 4th poorest regional country with a GDP fall of 8.6% (Comisión Económica para América Latina y el Caribe 2020; Consejo Nacional de Evaluación de la Política de Desarrollo Social 2018). As a consequence, high residential density and large population make cities most vulnerable to infectious disease outbreaks (Ghosh et al. 2020; Matthew and McDonald 2006), especially in the case of Mexico cities as mentioned above.

CDW quantification is the first step for CDW management policies to promote circular economy as a trigger for economic recovery. Many CDW quantification methodologies depend on local goals and scenarios, which could vary according to population augment, legislation, planning and the AEC industry (Jin et al. 2019; Wu et al. 2014). CDW quantity and quality depend on construction systems and materials, building typologies, age and demolition techniques (Menegaki and Damigos 2018). Additionally, CDW management plans are mandatory in many countries, requiring CDW selection, collection and transport to treatment plants or landfills, controlled by audits (Kabirifar et al. 2020). The increment of the AEC industry activity in developed and developing countries motivates the comparison of CDW management in different scenarios such as the EU, USA and China, recommending the adoption of emergent technologies, onsite audits, government supervision, economic incentives, interaction with stakeholders, coordination between operative departments and directions towards a circular economy (Aslam et al. 2020).

This research reviews different CDW quantification methodologies (Table 2 in Appendix 1) that include material flow analysis, direct and indirect onsite and offsite CDW quantification, CDW track load counts to treatment plants or landfills, surveys to workers, recyclers and government officials to follow CDW from origin to end and executive project plans consulting among others.

For instance, Mercader et al. quantify CDW from 10 social housing blocks, following the work breakdown structure of the Andalusian Construction Cost Database (BCCA) and the Royal Decree 105/2008 about CDW management in Spain (Barón et al. 2017; Mercader-Moyano and Ramírez-de-Arellano-Agudo 2013; Ministerio de la Presidencia 2008). This methodology quantifies CDW as the portion of the resource material that become waste during the construction process by transformation coefficients that, unlike others, calculate CDW weight and volume, allowing to measure, for instance, soil swelling.

On the other hand, some authors use municipal statistical data based on demolition permissions combined with GIS data about building area, volume, age and typology (Kleemann et al. 2017). Building Information Modelling is a platform where CDW quantification methodologies proliferate, providing several data sources for CDW production and involving stakeholders, designers and decision-makers to plan from design and procurement documentation to construction and demolition stages (Akinade et al. 2018; Cheng and Ma 2013; Liu et al. 2019; Won and Cheng 2017; Xu et al. 2019). Mercader Moyano et al. develop a quantification method for construction materials and CDW measuring their embodied energy and carbon emissions from cradle to the end of the construction stage and calculating indicators based on the BCCA work breakdown structure (Mercader Moyano et al. 2019). Many other authors develop CDW quantification methods using BIM tools (Bakchan et al. 2019; Ge et al. 2017; Guerra et al. 2019) and GIS data (Li et al. 2020; Miatto et al. 2019; Park et al. 2014; Tanikawa and Hashimoto 2009). Other authors link BIM with life cycle assessment (LCA) for CDW quantification (Jalaei et al. 2019). Likewise, LCA has provided a versatile tool to assess CDW from a circular economy framework (Jiménez Rivero et al. 2016; Luciano et al. 2018).

In Mexico, CDW quantification follows a methodology that considers the construction material purchases with a CDW production coefficient $= 0.3$ m^3 m^{-2} and another one to convert volume to weight $= 1.5$ ton m^{-3}, referred to research from the USA and Europe. The composition of CDW consisted in the observation and measure of truckloads sent to a landfill in Mexico Federal District (Secretaría de Medio Ambiente y Recursos Naturales 2010). In the commitment of the environmental goals proposed by the Development National Plan 2013–2018, SEMARNAT and the Mexican Chamber of Construction Industry delivered a National Plan for CDW management to implement the Mexican Standard NOM-161-SEMARNAT-2011. The CDW quantification focused on the tasks that produced the highest volumes of CDW while giving general recommendations about the CDW measurement and onsite selection. This report estimated that 6.08 million tons generated in

2011 would reach 9.2 million tons in 2018, with a rise of 3.5% of the AEC industry activity. CDW composition consists of 39% of soil excavation, 25% of concrete, 24% of mixed CDW and 12% of other types (Cámara Mexicana de la Industria de la Construcción 2016; Procuraduría Federal de Protección al Ambiente 2014). However, the lack of effectiveness of this CDW quantification model shows the need to obtain real coefficients from CDW production and management, which would favour the CDW circular economy, increasing Mexican GDP and offering an exit to the post-COVID crisis from the AEC industry.

This paper aims to develop a CDW management methodology through the transformation coefficient calculation of each construction material that becomes CDW, followed by its environmental footprint to quantify them from the design stage. This methodology, applied to social housing in Mexico, was developed and successfully implemented by the authors, giving rise to municipal regulations associated with the Andalusian Construction Costs Database and BIM software in Spain. Since current coefficients used in Mexico have failed to succeed because they base on foreign models, this methodology adapted to the Mexican scenario delivers data from onsite CDW quantification and user surveys after auditing 61 single-family housing over five years. This work will be the first step to help CDW management policies and municipal regulations to predict CDW production and promote reusing and-or recycling to minimize raw materials consumption. Furthermore, the environmental footprint addition to the original Spanish methodology accomplishes the EU Circular Economy Action Plan requirements (European Commission 2020b).

2 Objective and Methodology

The methodology consists of three stages that satisfy the objectives of the main scope that was mentioned above. In the first stage, the definition of the housing conventional construction model (CCM) requires establishing the typology features, select a representative sample and quantify the material resources consumed in the construction process. In the second stage, the CDW quantification requires its identification and characterization and the calculation of the transformation coefficients from the onsite measurement and surveys. In the third stage, CDW embodied energy (EE) and CO2e emissions combined with CDW destination characterize the CDW to assess its environmental footprint (Fig. 2).

This methodology applies the Spanish CDW methodology, which uses the weighted transfer of measurement (WTM) to identify and quantify CDW generated from the consumed material resources of the social housing CCM studied over ten years and developed by Mercader-Moyano and Ramírez-de-Arellano-Agudo (2013). It has also provided the basis for new CDW municipal regulations in Andalusia and Madrid (Ramírez de Arellano Agudo 2014). In this way, this work adds new CDW transformation coefficients and environmental footprint assessment to favour the transition from a linear economy model to a circular one. Consequently, CDW can be reintroduced into the economic cycle as reused, recycled and by-products.

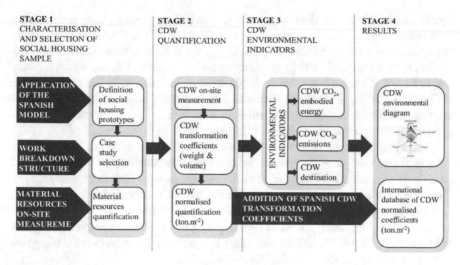

Fig. 2 Methodological framework and stages. *Source* The authors

3 Materials and Methods

The methodology is divided into three stages.

3.1 First Stage: Characterization and Selection of Social Housing Sample

Mexican social housing presents a built area between 42 and 76 m^2. On the other hand, the Building Code (CONAVI) states a lower size for them that includes a kitchen-dining room, one or two bedrooms, one bathroom, one parking space and basic services, considering that one family can satisfy all their needs there (Alderete Herrera 2010; Gobierno Federal de México 2016). Building materials and systems changed along with the evolution of the social housing in Mexico, so this work comprehends cases that are not more than 10-years-old to avoid referring to extinguished processes and materials. The CCM consists of reinforced concrete foundation slab and pillars, concrete block walls, reinforced concrete vault blocks and prefabricated beams, outer cement-based finish, inner plaster finish, ceramic floors, single-pane glazing with aluminium frame and wooden doors with aluminium frame (López-López 2019). In the '70s and '80s, several companies begin to build social housing neighbourhoods for low-income populations in urban outskirts; they comprehend one/two-story houses with a rigid scheme that do not respond to the household changing needs along their life cycle (Sánchez-Corral 2013).

The sample locates in Saltillo City, the capital district of Coahuila State, in Northern Mexico, next to the USA frontier. Despite its large volume of social housing neighbourhoods, the city government lacks the necessary data to determine CDW management policies (Hyman et al. 2015). In 2015, Coahuila State population accounted for more than 2,950,000 people, 27% of which (around 807,000 people) were distributed 90% in urban areas and 10% in rural areas, with a poverty rate of 61.1% (Consejo Nacional de Evaluación de la Política de Desarrollo Social (CONEVAL) 2020). Its strategic location, administrative importance and industrial development make this city a relevant data source in the social housing sector to approach CDW production in the AEC industry (Fig. 3).

Even when the social housing CCM changed traditional materials like adobe or quincha for modern ones like reinforced concrete, the construction techniques that local developers use are still artisan, as onsite surveys to supervisors and workers have shown (Figs. 4 and 5). During the last decade, the MCC morphology, construction systems and materials have been standardized by the different building companies in Saltillo, building nearly identical prototypes. The three main companies situated in SE Coahuila: DAVISA, RUBA and SERVER, produce the prototypes between 46 and 52.13 m^2 that are the basis for this investigation.

The BMCs list utilizes a material take-off (MTO) whose data sources are the Housing Institute of Coahuila State Government, local building companies and Saltillo housing developers. Due to different criteria among the Mexican States, this research adopts the average between 58 m^2 (from Coahuila State mortgages)

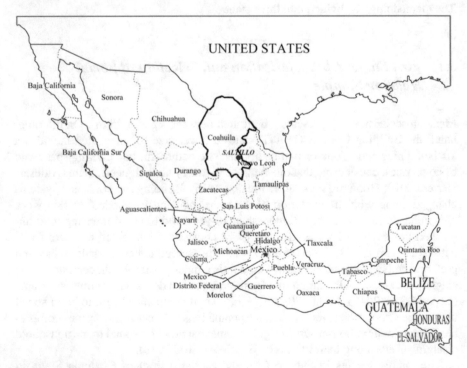

Fig. 3 Saltillo City location in Coahuila State, Mexico. *Source* Modified from López-López (2019)

Fig. 4 Social housing under construction. *Source* The authors

Fig. 5 Inner plaster finish and reinforced concrete vault block concrete slab. *Source* The authors

and 42 m^2 (from CONAVI prescriptions) that accounts for 49.4 m^2 with 90% confidence interval and 10% error in regard to the mortgages delivered in Saltillo during 2016 (López-López 2019) (Table 3 in Appendix 2).

The first step consists of a social housing historic and evolution approach to find the referential places in Saltillo City. After selecting the sample, the second step consists of the 61 social housing audits to identify and quantify the material resource consumption involving direct observation and surveys to workers and supervisors.

When following the model developed by Arellano (Ramírez de Arellano Agudo 2002), the work breakdown structure (WBS) codifies the CCM construction tasks. These data come from the executive project documentation and technical specification sheets. The bill of quantities (BOQ) provides the BMC consumption; each task code consists of one letter (X) and the addition of a sequence of two letters and three numbers in the case that there is more than one type (XYZ00n) while BMCs show a

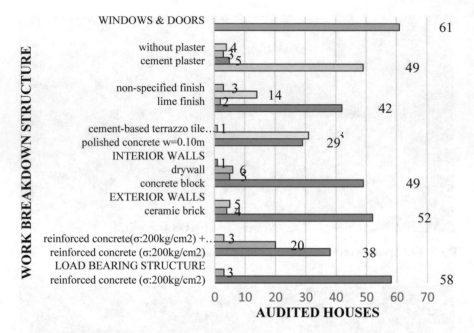

Fig. 6 CCM material resource consumption of the 61-house sample. *Source* The authors

3-letter code (TUV) (Table 4). The quantification of BMCs consumed in the sample construction uses kg m^{-2} units (Fig. 6; Table 5).

3.2 CDW Quantification

In this stage, the methodological steps let the CDW quantification through the transformation coefficients. From the onsite observation, it is possible to detect four steps of CDW production:

- Product delivery: deficient quality, download breaks, unproperly atmospheric conditions
- Storage: expired storage time, material breaks, package waste
- Construction tasks: internal transport breaks, mortar/concrete remains, material trimmings, badly executed works demolition, incorrect manoeuvre losses, machinery lubricant replacement, excavation soil not backfilled onsite
- Demolition waste.

After analysing different sources to address CDW materials and their codification like the Mexico City Government (GODF 2015), Mexican Federal Environmental Secretary (Medina Ross et al. 2001) and EU Waste Catalogue (Official Journal of the European Communities and Commission Decision 2001/118/EC of wastes 2001), it is decided to build an own classification and codification, according to Mexican regulations and following the systematic classification of the Andalusian Construction Costs Database (ACCD) as it has already shown its effectiveness since it was implemented (Barón et al. 2017). This codification allows identifying CDW from BCMs consumed in the construction process since they refer to the task, the type and material resource where CDW comes from (Table 1).

The transformation coefficient (CR) measures the proportional part of a BCM that becomes waste. This methodology called WTM, as mentioned in Sect. 2, is based on the onsite CDW quantification from the selected 61 houses construction sites (Fig. 7). This methodology successfully applied in Spain, provided the basis not only for the Spanish CDW management regulations but the EU 2030 Climate and Energy Framework as well (European Commission 2016).

CDW not only includes BCMs but other materials as well. Package (plastic, board, paper, tins and others) becomes CDW even though it is not a constituent part of the building. Some material resources undergo physical or chemical changes from their initial state after their installation or use and may vary their properties like ceramic CDW, which increases its volume and changes its use. Likewise, formworks, struts and scaffolds end their life cycle after a certain number of uses.

For the reasons explained above, it is necessary to apply indirect measurement methods to determine the quantities of CDW from the material resources, following the mathematic model (WTM) (Mercader-Moyano and Ramírez-de-Arellano-Agudo 2013) (Eq. 1).

$$Q_t = \sum_i^N Q_i \times CR_i \times CC_i \times CT_1 \tag{1}$$

where

Q_t　　CDW total quantity (ton)
N　　material resource/package index
Q_i　　quantity of material resource (ton)
CR_i　　transformation coefficient of the material resource that becomes CDW
CC_i　　CDW transformation coefficient from BCM unit to CDW unit
CT_i　　transformation coefficient from BCM measurement criterion to CDW criterion.

CR is directly measured at the construction place, while CC converts the material resource units to CDW units. Finally, CT converts the material resource measurement criterion to the CDW measurement criterion.

When applying Eq. 1 to this research, CC and CT coefficients are equal to 1 because CR is in tons. Packages CR is always equal to 1.

To compare CDW quantification of different typologies or locations, it is necessary to normalize them according to the corresponding built area (Eq. 2)

$$Q_{tn} = \frac{Q_t}{\text{built area}} \tag{2}$$

Table 1 WBS codification: tasks, task types, BCMs and CDW

Task		Task type					BCM				CDW			
Code	Description	Code	Description	Specifications	Quantity	Unit	Code	Description	Quantity	Unit	Code	Type	Quantity	Unit
X		XYZ00n					TUV				XYZ00n-TUV	BCM		
											XYZ00n-WWW	Pack		

Ref, CR, CC and CT: transformation coefficients
where
X = task alphabetical code
XYZ00n = task type alphanumeric code
TUV = BCM alphabetical code
XYZ00n-TUV = CDW alphanumeric code from BCM
XYZ00n-WWW = CDW alphanumeric code from BCM package
Source The authors

Fig. 7 CDW packages stored in the construction site. *Source* The authors

where

Q_{tn}	normalized CDW weight per area (ton m^{-2})
Q_t	CDW total weight (ton)
built area	built area of the building/s (m^2).

3.3 CDW Environmental Indicators

In Mexico, CDW may have four destinations: onsite reuse, onsite recycling, treatment plant for recycling and landfill. Therefore, CDW types are characterized and quantified according to their disposal to build four new CDW management indicators. Environmental Secretary of Mexico Federal District developed them in 2013 and updated them in 2015 (GODF 2015). This Standard compels the addition of reused or recycled materials to the executive project of a building if the second-use material is available within a radius of less than 20 km from the building site (Eq. 3).

$$T = RU + RC_o + RC_a + D \tag{3}$$

where

T	CDW total quantity (ton m^{-2})
RU	CDW onsite reuse (ton m^{-2})
RC_o	CDW onsite recycled (ton m^{-2})
RC_a	CDW recycled in treatment plant (ton m^{-2})
D	CDW sent to landfill (ton m^{-2}).

After CDW characterization and quantification according to its destination, the next step is to calculate two environmental footprint indicators according to CDW type: embodied energy and CO_{2e} emissions, whose sources are the BEDEC database and Arguello Mendez et al. research (Argüello Méndez and Cuchí Burgos 2008;

ITEC Instituto de la construcción de Catalunia n.d.) because Mexico lacks its own environmental footprint database.

Finally, a radial graphic synthesizes the CDW environmental indicators quantifying the environmental performance of an executive project. It intends to serve as a basis for an environment assessment labelling from a circular economy perspective. Furthermore, these indicators provide the quantity and destination of CDW to orient CDW management public policies to determine the need, location, size and type of treatment plants for reuse and recycle and landfills.

4   Results and Discussion

The proposed methodology allows measuring the CCM material resources (Table 5) and CDW produced at each stage of the building construction to obtain one CR coefficient for each BCM (Table 6). The implementation of a "hard" method like material tally during storage, installation, or construction process and the reckon of CDW bucket loads and truckloads complemented with a "soft" method of onsite surveys to workers and supervisors provide reliable data on real cases. The application of the coefficients (CR, CT and CC) accounts for the total normalized weight of the case study: 0.083 ton m^{-2}, without considering excavation soil (Table 7; Fig. 8). CDW painting (6.02E$-$2) does not appear because it constitutes hazardous waste.

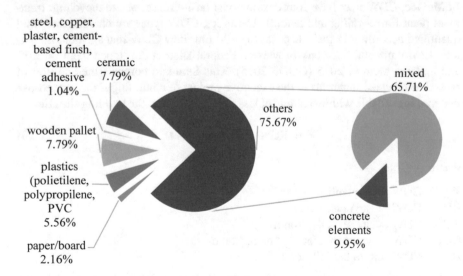

Note: Steel, copper, plaster, cement-based coating, and cement adhesive are grouped to facilitate the graphic reading just because they account for 1.04%.

Fig. 8 CDW percentages of the CCM. *Source* The authors

A Sankey diagram synthesizes the material flows from their arrival to the construction site as far as their destination as CDW (Fig. 9).

Concrete and cement-based materials account for the majority of the CCM resources provoking the largest amount of CDW; foundations, load bearing structure and walls. After adding mortars, they totalize 75.77% of CDW.

It can be observed that the CCM construction process is far from closing the material flow loop. The CCM consumes 1.24 ton m^{-2} and produces 0.083 ton m^{-2} CDW of which only a small amount is reintroduced as cement or lime packages. New construction materials only use raw materials, Mexican regulations compel neither manufacturers to add recycled steel or recycled inert aggregates nor constructors to use recycled materials if they are farther than 20 km from the construction site.

Notwithstanding, Mexican Federal District (F.D.) officials compel constructors to provide a CDW management plan coordinated with CDW transport service, in case that CDW overpasses $7 \, m^3$. The plan must include the authorized onsite storage places and recycling plants, or landfills. Nevertheless, the National CDW Management Plan estimated that only 20% of CDW from public and private constructions ended in authorized places, 77% in backfills, landfills, soil remediation and road subsoils and only 3% in recycling plants (Ambiente n.d.; Cámara Mexicana de la Industria de la Construcción 2016; CDMX 2019; Gobierno de México 2015; GODF 2015; Secretaría del medio ambiente 2018).

The CDW management plan proposed in this research would serve to organize the construction site leaving enough room at different times for CDW storage and inner transport during the construction, taking into account that foundations produce 19.78%, masonry, 64.69%, roof, 8.70% and finishes, 6.39% of total CDW while sanitary, drainage, electric and glazing systems account for 0.44% of the total CDW. It would also allow onsite CDW separation according to the destination: onsite reuse, recycling, recycling plant or landfill. In addition, CDW quantification let know the environmental footprint derived from the material resources employed in the CCM since the design stage. Hence, designers may substitute current construction materials for others with a lower environmental footprint.

Figure 10 shows the different CDW types per CCM square meter. Polyethylene packaging constitutes a special case among other plastics because it accounts for more than half of the CDW embodied energy while polypropylene and PVC account for minimal quantities (Fig. 11). Besides, it represents 71% of CDW CO_{2e} emissions (Fig. 12) and only 5% is recycled offsite (Fig. 13). Figure 14 shows the different CDW types sent to landfills (Table 7).

Then, a radial diagram synthesizes these data to characterize the executive project in regard to environmental indicators. They quantify CDW according to its destination: reused (RU), onsite recycle (RCo), offsite recycle (RCa) and landfill (D). RU, RCo and RCa show values equal to 0, or negative if CDW is recycled. Instead, D is always positive because it accounts for CDW which ends its lifecycle in a landfill. CDW embodied energy and carbon emissions appear on the two other axes. As it happens with building energy labelling, colours indicate CDW environmental indicators performance that means greater inefficiency from green to red. CDW sent to

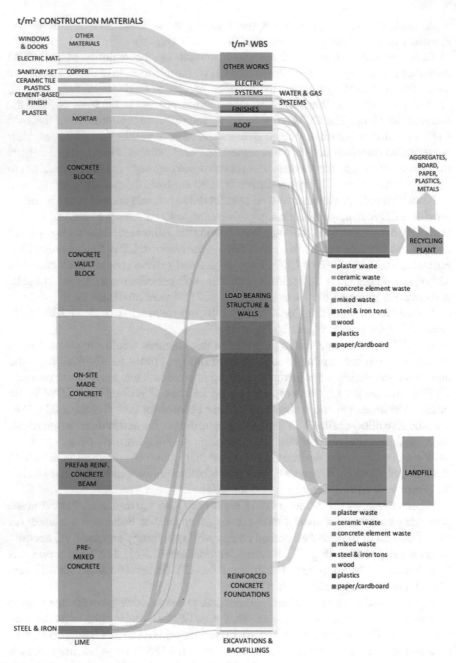

Fig. 9 Sankey diagram: CCM construction materials and CDW. *Source* The authors

Fig. 10 CDW weight
(ton m^{-2})

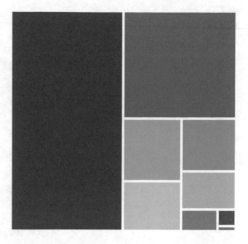

Fig. 11 CDW EE
(kWh m^{-2})

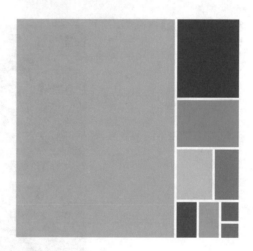

landfill (D) accounts for 69.20 kg m^{-2} while reused offsite CDW (RCa) comprehends 27.19 kg m^{-2} inert soil and only 19.92 kg m^{-2} other CDW (Fig. 15).

The proposed model validates its methodology when compared with Spanish social housing CCM that consumes 1.9 ton m^{-2} and produces 0.08 ton m^{-2} of CDW (Mercader-Moyano and Ramírez-de-Arellano-Agudo 2013). Both models omit inert soil as CDW. In the Mexican case, 6.71% of material resources become CDW, while it accounts for 4.21% in the Spanish case, deducing a more efficient CDW management for this last one. Nevertheless, mixed CDW accounts for 6.27E−2 ton m^{-2} (75.76% CDW) in the Mexican CCM and 6.80E−2 ton m^{-2} (85.13% CDW) in the Spanish CCM. After adding mixed CWD with ceramic, the total amount rises to 83.44% in the Mexican case and 95.61% in the Spanish case (Fig. 16; Table 8).

Results show that mixed CDW (mortar, onsite made concrete and concrete elements) accounts for the largest volume and weight. This waste could be grinded

Fig. 12 CDW CO$_{2e}$
(ton m^{-2})

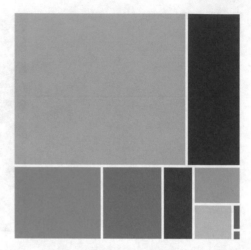

Fig. 13 Recycled CDW
(ton m^{-2})

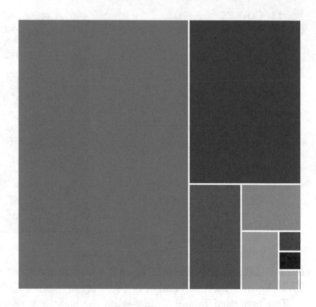

onsite to use as inert aggregate for subsoils ("Concretos Reciclados" n.d.) or to make onsite blocks as it happened in an exceptional case reported in Buenos Aires (Yajnes et al. 2017). Nonetheless, many authors propose to use it as inert aggregate in prefabricated blocks (Luciano et al. 2020; Pacheco-Torgal 2014; Rakhshan et al. 2020). These experimental materials represent a solution to make concrete blocks, but they still lack the required certification, dissemination and acceptance from users, developers and stakeholders, as it happens with blocks with added cardboard or just adobe bricks (Molar-Orozco et al. 2020; Roux Gutierrez et al. 2015).

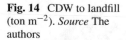

Fig. 14 CDW to landfill (ton m^{-2}). *Source* The authors

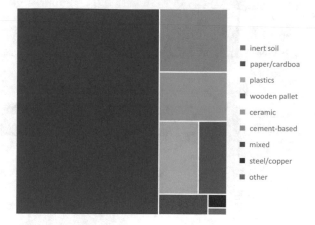

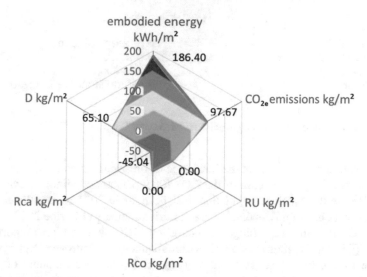

Fig. 15 CDW environmental indicators of the CCM. *Source* The authors

As it is observed, steel can be recycled reducing its environmental footprint, recovering part of their EE and the carbon emissions from its production process. However, recycling cement-based products partially reduce the EI and carbon emissions of cement production because their by-products present low aggregate value.

As this work has pointed out, Mexico still lacks mandatory CDW quantification and environmental footprint assessment, so the proposed methodology is determinant to implement federal and state legislation for CDW recovery to reintroduce them in the productive chain, minimizing those with low aggregated value that end in infrastructure work subsoil but mostly in landfills. These latter constitute pollution and infection focus exposing vulnerable populations that live in the surroundings.

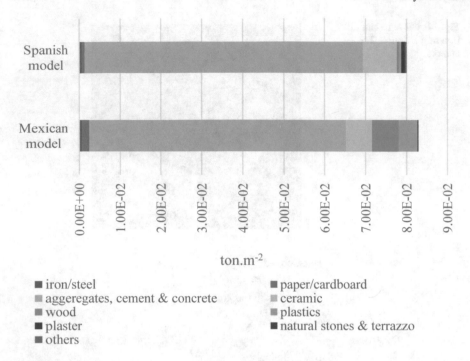

Fig. 16 CDW per type: Mexican and Spanish CCM. *Source* The authors

Moreover, landfills augment the incidence of breathing illnesses, increasing risk groups comorbidity in the case of the COVID19 pandemic, without leaving aside that 66.9% of the Mexican population will be poor in 2021 as mentioned in the Introduction (Comisión Económica para América Latina y el Caribe 2020).

Likewise, landfills emit biogases, including GHG with a high proportion of methane (CH_4) and carbon dioxide (CO_2), small amounts of nitrogen (N_2), hydrogen sulphide (H_2S), hydrogen (H_2) and oxygen (O_2) and trace amounts of carbon monoxide (CO), ammonia (NH_3), aromatic and cyclic hydrocarbons and volatile organic compounds (VOC). Environmental issues caused by these gases comprehend from nauseating odours to Global Warming contribution. Moreover, some of them as NH_4, CO, VOC and CO_2 have a direct harmful effect on human health. Furthermore, refrigerant and foaming gases and aerosols like chlorofluorocarbons (CFC), hydrochlorofluorocarbons (HCFC), hydrofluorocarbons (HFC) and halons provoke atmospheric ozone (O3) depletion. Besides, landfill leachate may pollute soils and superficial/subterranean adjacent water bodies, causing toxicity, eutrophication and acidification issues. Landfills also cause the proliferation of fauna that transmit illnesses or act as their vector like flies, mosquitoes, fleas, rats, birds and others that provoke diarrhoea, typhus, paludism, giardiasis, dengue fever, bubonic plague, leptospirosis and toxoplasmosis among others, rising public health costs (Gobierno de México 2015).

5 Conclusions

This work proposes a methodology for CDW quantification and environmental characterization applied to 61 social housing in Saltillo City, Coahuila, Mexico. This methodology considers not only CDW weight and volume but CDW physical or chemical transformation and the changes in their measurements criteria. It validates its results by contrasting CDW quantification at the executive project stage with the CDW amounts effectively measured onsite along the construction process.

The obtained 38 CR coefficients are added to the database developed in the CDW quantification methodology that has served as a basis for this research (Mercader-Moyano and Ramírez-de-Arellano-Agudo 2013). After being contrasted with the Spanish CR coefficients, it is possible to apply the new ones to the Mexican CCM to foresee CDW types and quantities and to plan its storage onsite and CDW destination for reuse or recycle. This methodology may serve as a basis for the implementation of local and national regulations.

Regarding the Spanish methodology, this research not only validates its application to any other country but updates and complements it with the addition of a multi-dimensional environmental footprint indicator. The design of this indicator synthesizes CDW environmental footprint and destination for each executive project in a radial graphic. Moreover, it may be a precedent for CDW environmental assessment labelling.

On the other hand, Mexico lacks enough CDW selection plants and authorized landfills favouring unauthorized places. Additionally, the lack of control of mandatory regulations makes that a high proportion of CDW ends in the mentioned illegal places, increasing environmental risks (Turcott Cervantes et al. 2021). These barriers require an integrated plan that covers the needed policies to shift from a linear economic model of extraction, use and discard towards a circular one of transformation of waste into a resource, involving all those AEC industry actors, generating green jobs, environmental benefits and motorizing the weak economy after the COVID19 pandemic. Notwithstanding, the authors find some limitations in this research: one is the dependence on staff collaboration for surveys. Another one is CDW destination verification because of the many illegal landfills where CDW may end. To solve the first one, a kind of incentive could improve the collaboration and for the second one, strict monitoring could assure the result accuracy.

These updates and additions prepare this new method to fulfil the new exigencies and objectives of the EU Circular Economy Action Plan that Spain had not accomplished by 2020 (European Commission 2020b).

Possible future lines of research will orient to the addition of CR coefficients to the CDW normalized database for social housing, by extending the work to other building typologies over the whole Mexican territory.

Other research lines could be the CDW economic valorisation to implement business models that make its transformation viable after the pandemic economic crisis. Circular business models involve local resources, reduce import dependence on supplies making the AEC industry more resilient and as they are labour intensive, they can create new jobs: EU estimations calculate around 600,000/700,000 net jobs by 2030.

Other lines of research may be the real estimation costs of CDW management developed from the executive project stage to include them in social housing public procurements through a BIM model as proposed by Mercader-Moyano et al. (2019). The addition of a CDW management plan and a fee to the municipal permission would compel the promoter to commit to these new regulations.

One of the CC and pandemic consequences will be the worsening of the already existing poverty and inequality in countries like Mexico (López-Feldman 2014).

A proper CDW management as part of a circular economy model delivers undoubtful benefits to the environment reducing the increment of GHG emissions, contributing to Climate Change (CC) mitigation while eradicating open-air garbage dumps to limit environmental pollution. In the EU, the circular transition could reduce 300 million tons per year by 2050. Reuse and recycling may produce second-use materials that close the material flow loop and diminish the depletion of raw materials. Also, CDW management may deliver social benefits like sanitary conditions and public health improvement, construction and operation of new facilities create employment and workers training improves quality labour, especially in local communities favouring social inclusion and diminishing poverty.

Nevertheless, the fall of raw material prices, the disruption of supply chains and the reduced investment that worsens liquidity might threaten the advances towards a circular model. Notwithstanding, the COVID-19 recovery would foster resilience, sustainability and inclusion if labour and capital resources underpin the circular economy transition agenda like recycling infrastructure instead of remaining idle in a low aggregate demand economy. This model improves the capital assets productivity, efficient production of materials and waste reduction expanding the resource base of the economy.

In the EU, approximately 15% of construction materials become waste during the construction phase showing a 50% overuse of steel and concrete materials. Likewise, aluminium recycling market could augment from EUR 3 billion to EUR 12 billion with a recycling rate increment by 2050. About 180/190 million tons of steel, plastics and aluminium lose their original value of EUR 140/150 billion when they reach their end of life preserving only 41% each year through volume or price losses. If the recycling process could be improved in quality and quantity, the EU could gain new resources, independence from imports, new jobs and help meet its CC goals. Better CDW waste management and recycling technologies and design for recyclability may maintain the material value longer in the economic cycle (Klevnäs and Kulldorff 2020).

Acknowledgements This paper is funded by "Mapei Spain S.A." for financial support via the research contract (68/83) ref.: 3719/0632 "Implementation of eco-efficient and urban health measures in building renovation and urban regeneration for the requirement of new products in the construction sector—2nd phase" and a Research Personnel Training Grant (arts. 68/83 LOU) ref. 3719/0632. This research was also developed due to the financial support of the University of Seville (VI PPIT-US; 2.2.3. 2018 grants) for the research stay of Jesús López-López at University of Seville in Spain. Finally, certain data and procedures from this work were gathered during the doctoral research developed by the author entitled "Characterization of CDW in Mexico. A theoretical model", in co-supervision between the University of Seville and the Autonomous University of Tamaulipas.

Appendix 1

See Table 2.

Appendix 2

See Tables 3, 4, 5, 6, 7 and 8.

Table 2 Comparison of CDW quantification methods

Author	Country	Study case	CDW quantification methodology	Conclusions
Villoria Sáez et al. (2018)	Spain	Residential façade renovation in Jerez and Madrid cities	Visits to construction sites for direct and indirect measurement	Focused on the reduction of concrete, ceramic, metal and wood CDW
Blaisi (2019)	Saudi Arabia	CDW in the Eastern Province, Saudi Arabia	Truck loads measurement from construction sites to landfills	Lack of institutional collaboration, CDW management policies, coordination among the different actors, incentives, mandatory laws for CDW collectors, landfills and recycling plants
Kartam et al. (2004)	Kuwait	Kuwait's recycling plants	CDW minimization	CDW recycling plants: advantages and disadvantages
Cha et al. (2020)	South Korea	1034 residential buildings	CDW rate measurement before demolition to calculate carbon emissions and potential recyclability according to economic and environmental values	Plastic waste has the highest recycling rate followed by wood, aggregates and metals

(continued)

Table 2 (continued)

Author	Country	Study case	CDW quantification methodology	Conclusions
Wu et al. (2016)	China	Shenzhen city	Deep surveys to recyclers and officials to gather data on CDW in construction and demolition sites from CDW production to final disposition. Extrapolation and CDW production methods, material flows and possible CDW use	CDW production is increasing in Shenzhen city so higher recycling rates may augment CDW economic potential
Yuan (2017)			Data collection through publications, government regulations, surveys and group discussions with public authorities and industry actors	Lack of coordination among the government departments, reliable data, construction project supervision and slow increment of recycling plants
Li et al. (2013)		New residential building at Shenzhen city	CDW production index based on built area, bill of quantities and material waste rate	Coefficient table for CDW generation for AEC industry
Jaillon et al. (2009)		Hong Kong	Potential CDW reduction by implementation of prefabricated systems	
Wu et al. (2019)			Critical comparison of CDW quantification methods in different scenarios	Detailed pros and cons of them
Li et al. (2016)		High-rise building in Tangshan	CDW quantification from the work breakdown structure and bill of quantities	CDW calculation at design phase to help developers minimize CDW

(continued)

Table 2 (continued)

Author	Country	Study case	CDW quantification methodology	Conclusions
Kleemann et al. (2017)	Austria	Vienna	Statistical analysis based on demolition permissions combined with GIS data: construction year, typology, area, volume to calculate gross demolition volume and transform them into different material rates: kg/m^3	Underestimation of CDW quantities because of statistical data that distort the following management
Barón et al. (2017)	Spain	20 housing blocks at design stage in Andalusia	CDW management costs added to the Andalusian construction cost database following the European waste catalogue	CDW codification, measurement and valorisation criteria
Mercader-Moyano and Ramírez-de-Arellano-Agudo (2013)	Spain	10 social housing blocks in Seville	aplicación de coeficientes de transformación a los materiales que, a diferencia del resto de las metodologías analizadas, permiten medir los residuos en peso y volumen	Flexible method to be applied to other countries
Ram and Kalidindi (2017)	India	Demolition of 45 buildings	CDW quantification rates according to typology and demolition area	CDW accurate estimation to implement management policies
de Guzmán Báez et al. (2012)	Spain	Railway works	CDW predictive model by work site selection, bill of quantities and CDW onsite quantification	CDW quantification in weight and volume units

(continued)

Table 2 (continued)

Author	Country	Study case	CDW quantification methodology	Conclusions
Carpio et al. (2016)		Housing blocks in Granada city	CDW quantification in different urban planning scenarios	CDW quantity depends on different typologies from the urbanization stage
Parisi Kern et al. (2015)	Brazil	16 high-rise housing buildings with different design floors	Floor plan influence on CDW quantification using multiple regression	Need to identify and consider all the variables that were considered in the design and affect the CDW production
Ajayi et al. (2015)	UK	Theoretical model	BIM and LCA	CDW quantity reduction through design executive project stages considering building material standard measures and implementing construction methods that minimize waste from broken pieces, excess and other causes
Jalaei et al. (2019)	Canada	12-story residential building in Montreal		BIM/LCA joint to reduce CDW and environmental footprint
Jiménez Rivero et al. (2016)	Spain	Recycling of gypsum board in the EU-27	Gypsum board LCA	Gypsum board recycling reduces certain aspects of environmental footprint while increases others depending on possible scenarios

(continued)

Table 2 (continued)

Author	Country	Study case	CDW quantification methodology	Conclusions
Luciano et al. (2018)	Italy	Public works	BIM and LCA: construction materials and CDW quantification	DECORUM integrates production phases with efficient resource use and CDW
Akinade et al. (2018)	UK	Surveys to AEC industry manufacturers	BIM	The factors that affect this sector are BIM-based collaboration for CDW management, CDW-based design, building LCA, innovative technologies for CDW management
Cheng and Ma (2013)	China	47-story residential building	BIM applied to construction, demolition and renovation waste (CDRW)	BIM-CDRW integration to calculate transport and disposition rate
Liu et al. (2015)	UK	100 surveys to architect firms	BIM framework for CDW minimization from design stage	Possible extension to other AEC sectors
Liu et al. (2019)	China	2 single-family houses	BIM optimization of roof sheathing	Efficient use of construction material through automated cut

(continued)

Table 2 (continued)

Author	Country	Study case	CDW quantification methodology	Conclusions
Won and Cheng (2017)	Several countries	Several study cases	BIM methodology analysis	Identification of limitations according to building construction processes, technologies, limitations identification, policies and BIM applied to CDW management
Wu et al. (2019)	Several countries	Several study cases	Assessment of CDW quantification methodologies: CDW rate, onsite visits LCA	Selection of the correct CDW methodology according to contextual scenario
Mercader Moyano et al. (2019)	Spain	Social housing blocks	Transformation coefficients applied to CDW embodied energy and carbon emissions embedded in EIM according to work breakdown structure of the Andalusian construction costs database	Normalized quantification method applicable to different cities and countries
Ge et al. (2017)	Australia	Building in university campus	BIM applied to DW	CDW management strategies differ from construction ones; reuse and recycling rates improvement and demolition and logistic cost reduction
Guerra et al. (2019)	EEUU	4000 m^2 building in university campus	CDW management through BIM bill of quantities	This methodology improves efficiency in CDW management

(continued)

Table 2 (continued)

Author	Country	Study case	CDW quantification methodology	Conclusions
Bakchan et al. (2019)	EEUU	40,135 m² pilot building in university campus	CDW calculated in BIM and contrasted with a pilot case	Difference of 5.3% between estimated and measured CDW
Zaman and Lehmann (2013)	Australia, EEUU and Sweden	Adelaide, San Francisco and Stockholm	"Zero Waste Index" measure the material recovery capacity within CDW flows quantifying raw material, energy and carbon emission savings	This tool assesses CDW management and material substitution through management systems applied to different cities
Li et al. (2020)	Several countries	Several study cases	BIM, GIS	Digital tools prevalence over traditional ones for CDW management
Tanikawa and Hashimoto (2009)	UK-Japan	Urban areas in Salford, UK and Wakayama, Japan	4D GIS	Building stock renovation at urban scale includes underground infrastructure within CDW
Park et al. (2014)	South Korea	Public procurement case	CDW quantification of 7 from 52 construction materials in BIM	Aplicación de la herramienta para licitaciones públicas
Miatto et al. (2019)	Italy	Padova (1902/2007)	CDW material flow through map and aerial photographs analysis of building renovation promoted by public policies on energy efficiency and asbestos removal	GIS-based data of building stock to encourage public policies on CDW management

(continued)

Table 2 (continued)

Author	Country	Study case	CDW quantification methodology	Conclusions
Lu et al. (2015)	China	5764 buildings in Hong Kong	Big Data: CDW production rate	Results are reliable only as benchmarks
Cochran and Townsend (2010)	EEUU	EEUU	CDW material flow quantification	Methodology with potential application to other countries
Ding and Xiao (2014)	China	Shanghai	CDW material flow quantification by weight/area	Concrete, bricks and blocks constitute more than 80% CDW. 50% of them could be recycled
Bakchan et al. (2019)	Lebanon	CDW quantification of 28 construction sites in Beirut	CDW material flow quantification	CDW amounts 38–43 kg/m^2, masonry and concrete accounts for 60% of them

Table 3 Building elements composition of the 61 audited social housing

Task	Construction system	Number of dwellings	Total
Load bearing structure	Reinforced concrete (σ: 200 kg/cm^2)	58	61
	Reinforced concrete (σ: 200 kg/cm^2) + steel/wood	3	
Roofs	Reinforced concrete (σ: 200 kg/cm^2)	38	61
	Lightened reinforced concrete with EPS blocks	20	
	Reinforced concrete (σ: 200 kg/cm^2) + steel/wood	3	
External walls	Concrete block	52	61
	Ceramic brick	4	
	Concrete block + brick	5	
Internal walls	Concrete block	49	61
	Ceramic brick	5	
	Drywall	6	
	Concrete block + drywall	1	
Floors	Polished cement 0.10 m	29	61
	Vitrified tile	31	
	Cement-based terrazzo 0.30 × 0.30 × 0.025	1	
External coatings	Cement-based	42	61
	Lime	2	
	Premixed	14	
	Non-specified	3	
Internal coatings	Plaster	49	61
	Cement-based	5	
	Non-specified	3	
	No coating	4	
Windows/doors	Inner wooden doors + Al frame		61
	Steel cover door + wooden frame		61
	Kitchen steel door + steel frame		61
	Single-glazed Al windows		61

Table 4 Codes according to CCM social housing tasks and material resources

Task	Task type		Material resources		
	Code	Description	Family	Code	Description
A. Excavations and backfillings	ADS	Plot clearing	N. Raw material	NTI	Inert soil
	ALT	Plot cleaning		NTO	Organic soil
	ALD	Plot clearing and cleaning		NAA	Sand/gravel
	ANC	Plot levelling and compaction		NPN	Natural stone
	ANV	Plot levelling		NCL	Lime
	ACT	Plot compaction	H. Cement-based	HHC	Cement
	ATR	Plot demarcation		HHO	Onsite made concrete
	AEC	Foundation excavation		HHP	Premixed concrete
	AEP	Well excavation		HMM	Mortar
	AEN	Levelling excavation		HBC	Concrete block
	ARE	Backfilling		HPA	Artificial stone
	ARN	Levelling landfilling		HMP	Cement-based terrazzo
	AWW	Other works		HRI	Exterior coating
B. Foundations	BMM	Stone masonry secured with cement mortar		HAI	Adhesive
	BCC	Cyclopean concrete: 40% stone min. $\Phi 10''$		HEI	Joint filler
	BZA	Reinforced concrete basement		HVI	Prefabricated reinforced concrete beam
	BZC	Reinforced concrete continuous basement		HBO	Concrete slab block
	BLC	Reinforced concrete plate slab	M. Metals	MHC	Steel
	BDC	Foundation beam		MAA	Aluminium
	BME	Levelling wall		MAB	Bronze
	BMC	Containment wall		MBC	Copper
	BTL	League beam		MBL	Brass

(continued)

Table 4 (continued)

Task	Task type		Material resources		
	Code	Description	Family	Code	Description
	BCE	Especial foundation		MHH	Iron
	BWW	Other works		MCC	Annealed wire
C. Masonry	CKH	Reinforced concrete pillars		MPP	Lead
	CKA	Reinforced concrete column		MRR	Mixed waste
	CCE	Reinforced concrete top ring beam		MOO	Insulation material
	CVI	Reinforced concrete beam	Y. Gypsum-based	YYY	Gypsum
	CCM	Reinforced concrete slab	L. Cellulose	LMD	Wood
	CAB	Reinforced concrete slab with ceramic brick vault		LMP	Industrialized wood
	CVC	Reinforced concrete slab with concrete block vault		LPC	Paper/cardboard
	CVP	Reinforced concrete slab with EPS block vault		LTM	Wooden pallet
	CBA	Bare concrete block wall	P. Industrialized metals	PHV	Reinforcing steel bar
	CLA	Bare ceramic brick wall		PHR	Electric wire
	CCA	Bare reinforced concrete wall		PME	Steel bar mesh
	CMB	Concrete block wall		PCL	Nail
	CMP	Stone wall		PKC	Reinforced steel bar column
	CML	Ceramic brick wall	C. Ceramic	CPC	Ceramic floor
	CMC	Reinforced concrete wall		CAZ	Tile
	CMW	Other walls		CMC	Walls
	CPC	Cement mortar	A. Clay-based	ACT	Roof tile

(continued)

Table 4 (continued)

Task	Task type		Material resources		
	Code	Description	Family	Code	Description
	CFC	Lean concrete subfloor		ALD	Brick
	CFA	Reinforced concrete subfloor		ABB	Adobe brick
	CBC	Concrete floor		ABA	Lightened brick
	CRP	Concrete moulding	S. Synthetic	SMP	Plastics
	CRS	Drainage chamber		SMS	Synthetics
	CWW	Other works		SPV	PVC
D. Roof	DIM	Waterproof coating	T. Painting	TBA	Water-based
	DPB	Concrete block parapet		TSE	Sealer
	DPL	Ceramic brick parapet			
	DCH	Concrete chamber			
	DBT	Water tank base			
	DCP	Rainwater basin			
	DBP	Rainwater drain pipe			
	DPP	Tilted subfloor			
	DWW	Other works			
E. Coatings	EAP	Cement-based plaster			
	EAI	Industrialized cement-based plaster			
	EYS	Gypsum plaster			
	ELC	Ceramic tile coating			
	EPC	Ceramic tile floor			
	EPB	Mud floor			
	EPF	Polished cement floor			
	EZB	Mud baseboard			
	EZC	Ceramic baseboard			
	ETX	One-layer textured coating			
	EPV	Vinylic painting			

(continued)

Table 4 (continued)

Task	Task type		Material resources		
	Code	Description	Family	Code	Description
	EWW	Other coatings			
F. Water, sewer and gas systems	F1H	Water provision			
	F2S	Sewer			
	F3G	Gas			
	F5W	Others			
G. Electric, lighting and communication systems	G1E	Electric			
	G2I	Lighting			
	G3T	Communication			
	G4D	Automation			
	G5W	Others			
H. Windows, doors	H1A	Aluminium			
	H2M	Wood			
	H3A	Steel			
	H4C	Hinges, latches, locks			
	H5P	Plastic			
	H6W	Others			
I. Outside works	I1J	Gardening and landscape			
	I2I	Outer lighting			
	I3O	Complementary works			
	I4L	Final cleaning			
	I5W	Other works			

Table 5 Material resources quantification of social housing

Task		Task type					Material resources					
Code	Description	Code	Description	Specifications	Q	u	Code	Description	Deliver. quant.	Net quant.	Orig. unit	Normal quant. kg/m²
A	Soil excavation-prelimin. works	ATR001	Demarcation		1	pl	NCL	Lime	5.00E-03	5.00E-03	t	1.01E-01
B	Foundation	BLC001	Reinforced concrete foundation slab	Premixed-concrete Holcim ($\sigma'c = 200$ kg/cm²), coarse aggr. max. Φ 3/4″, slump test: 0.12 cm, steel bar, double mesh Φ3.8″ (0.20 × 0.20 m), superimposed in both directions, tied with annealed wire (BWG 16)	49.40	m²	HHP200	Premixed concrete	6	5.93	m³	2.88E+02
							PHV3-8	Steel bar	595	561	kg	1.14E+01
							PHVC-16	Annealed wire	4	3.84	kg	7.77E-02
C	Masonry	CKH001	Reinforced concrete pillar	Onsite made concrete ($\sigma'c = 150$ kg/cm²), coarse aggr. max. Φ 3/4″, casted in concrete block wall (6″ width), 1 steel bar Φ 3.8″ fixed to foundation slab (L-shape), 0.40 m length, separated 1.20 m each	102.45	m	HHO150	Onsite concrete	1.91	1.75	m³	8.50E+01
							PHV3-8	Steel bar	66.84	64.83	kg	1.14E+01
		CCE001	Top ring beam	Concrete ($\sigma'c = 150$ kg/cm²) 0.15 × 0.20 m, prefabricated steel frame ARMEX DeAcero™ 0.12 × 0.18 m	12.75	m	HHO150	Onsite made concrete	0.42	0.38	m³	1.86E+01
							PKC12	Prefab. frame ARMEX 0.12 × 0.18 m	13	12.75	m	4.23E-01
		CVC001	Reinforced concrete slab with concrete block vault	Onsite made concrete 0.15 high compression layer 0.04 m high steel mesh DeAcero™ 66-1010	49.40	m²	HVI	Prefab. reinforced concrete beam	66	65.87	m	7.20E+01
							HBO	Concrete block	268.57	263.30	Block	2.16E+02
							PME	Steel mesh	50	49.40	m²	5.00E+00

(continued)

Table 5 (continued)

Task		Task type					Material resources					
Code	Description	Code	Description	Specifications	Q	u	Code	Description	Deliver. quant.	Net quant.	Orig. unit	Normal quant. kg/m²
		CMB001	Concrete block wall	Concrete block 0.15 × 0.20 × 0.40 m laid with cement mortar 1:5	102.56	m²	HHO150	Onsite made concrete	2.16	1.98	m³	9.60E+01
							HBC15	Concrete block 0.15 × 0.20 × 0.40 m	1400	1370	Block	1.66E+02
							HMM1-5	Onsite made mortar (1:5)		0.92	m³	3.35E+01
		CBC001	Concrete floor auxiliary yard	$\sigma'c = 150$ kg/cm², 0.08 m width polished	5.16	m²	HHO150	Onsite made concrete		0.41	m³	6.00E+01
		CBC002	Concrete floor parking space	$\sigma'c = 150$ kg/cm², 0.08 m width textured	15.44	m²	HHO150	Onsite made concrete		1.24	m³	6.00E+01
		CBC003	Concrete shoulder	$\sigma'c = 150$ kg/cm², 0.08 m width, squared textured 0.65 × 1.20 m	7.80	m²	HHO150	Onsite made concrete		0.62	m³	3.03E+01
		CRS	Sewer chamber	$\sigma'c = 75$ kg/cm², floor width = 0.08 m polished, concrete walls 1.10 × 0.20 × 0.40 m onsite made, reinforced concrete cover $\sigma'c =$ 100 kg/cm² steel mesh 66-1010	3	u	HBC15	Concrete block 0.15 × 0.20 × 0.40 m	9	8.01	Block	9.73E-01
							HMM1-5	Onsite made mortar (1:5)		4.03E-04	m³	1.47E-02
							HHO75	Onsite made concrete		0.04	m³	1.71E+00
							HHO100	Onsite made concrete cover		0.02	m³	1.07E+00
							PME	Steel mesh 66-1010		0.44	m²	4.45E-02
D	Roof	DIM001	Insulation/waterproof coating	Elastomeric acrylic coating Thermotek™ Max 7, type 7ᵃ, cold aqueous emulsion, reinforced with simple membrane Thermotek™	59.30	m²	PBA	Insulating/waterproof painting × 200 l	62.56	59.30	1	1.68E+00
							SMS	Simple membrane Thermotek™ × roll = 109.25 m	62.56	59.30	m²	4.20E+00

(continued)

Table 5 (continued)

Task Code	Task Description	Task type Code	Description	Specifications	Q	u	Material resources Code	Description	Deliver. quant.	Net quant.	Orig. unit	Normal quant. kg/m²
		DPB001	Concrete block parapet	Onsite made with mortar 1:5 and on top mortar chamfer (0.05 a 0.01 m)	33	m	HBC15	Concrete block 0.15 × 0.20 × 0.40 m	82.10	80.49	Block	9.78E+00
							HMM1-5	Onsite made mortar 1:5		0.05	m³	1.97E+00
		DCH001	Concrete chamfer	Mortar 1:5, 45°	32	m	HMM1-5	Onsite made mortar 1:5		0.36	m³	1.31E+01
		DBT001	Storage water tank base	h:2 concrete block rows laid with mortar 1:5 and onsite made reinforced concrete plate w = 0.08 m steel mesh 66-1010 DeAcero™ plastered with stucco Cemix™	1	u	HBC15	Concrete block	10.90	10.69	Block	1.30E+00
							HMM1-5	Onsite made mortar 1:5		0.01	m³	2.61E-01
							HHO100	Onsite made concrete		0.08	m³	3.89E+00
							PME	Steel mesh 66-1010	1.003	1.00	m²	1.01E-01
							HMMP	Stucco Cemix	17.23	16.3	kg	3.30E-01
		DCP	Rainfall drainage pipe	PVC pipe Φ2'/	2	u	SPV	PVC pipe Φ2'/	6.00	5.90	m	4.78E-01
E	Coatings	EAI001	Premixed mortar coating	Exterior block coating Aderblock CEMIX™ w = 0.005 m max. on primary sealer 5-1 CEMIX™	69.80	m²	HMMP	Stucco Cemix	0.07	0.07	t	1.41E+00
							HRI	Primary sealer	24.43	23.27	kg	4.71E-01
		EYS001	Gypsum coating	Width = 0.015 m	151.25	m²	YYY	Gypsum		0.61	kg	1.22E-02
		EYS002	Ceiling gypsum plastering	w = 0.015 m	42.70	m²	YYY	Gypsum		170.80	kg	3.46E+00
		ELC001	Interior ceramic coating	White enamelled ceramic Tile grade 1A model Alcalá Vitromex™	7.62	m²	CMC	Ceramic	8.08	7.62	m²	1.39E+00
							HAI	Adhesive Cemix™	2.96	2.82	kg	5.71E-02
		ELC002	Ceramic tile coating on façades	Ceramic tiles DalGres, Slate Brown 1°, Daltile™ 0.605 × 0.605 m laid with floor adhesive Pegamix Constructor Cemix™	4.80	m²	CMC	Ceramic tile	5.09	4.80	m²	8.74E-01
							HAI	Adhesive Cemix™	1.87	1.78	kg	3.60E-02

(continued)

Table 5 (continued)

Task Code	Description	Task type Code	Description	Specifications	Q	u	Material resources Code	Description	Deliver. quant.	Net quant.	Orig. unit	Normal quant. kg/m²
		EPC001	Ceramic tile interior floor	Ceramic tiles DalGres, Slate Brown 1°, Daltile™ 0.605 × 0.605 m laid with floor adhesive Pegamix Constructor Cemix™	41.95	m²	CMC	Ceramic tile	44.47	41.95	m²	7.64E+00
							HAI	Adhesive Cemix™	16.31	15.54	kg	3.15E−01
		EZC001	Ceramic tile baseboard	Ceramic tiles DalGres, Slate Brown 1°, Daltile™ 0.605 × 0.605 m laid with floor adhesive Pegamix Constructor Cemix™	43.75	m	CMC	Ceramic tile	5.57	5.25	m²	9.56E−01
							HAI	Adhesive Cemix™	0.10	1.94	kg	3.94E−02
F	Water, sewer and gas systems	F1H001	Water pipes	Thermally fused PP pipe Random/PP-R (NMX-E-2262-CNCP-2007) Tuboplus, Rotoplas™, Φ3/4″-1/2″ within foundation slab and walls	45.00	m	SMP	PP pipe thermally fused	45.90	45.00	m	7.29E+00
		F1H002	Shower set	2 ceramic lock mixing faucets, shower with sprinkler, 2 PVD brushed nickel-plated stopcocks, model Builders Glacier Bay™	1	u						6.07E−02
		F1H003	Bath set	Bath set HD Cosmos II RD 4 pieces WC white with rounded storage tank 4.8 l with frontal handle, basin and washstand Orion™	1	u						9.78E−01
		F1H004	Laundry set	Granite worktop with laundry basin 65 × 50 cm	1	u						1.12E+00
		F2S001	Sewer piping	PVC pipe (NMX-E-215/1-1994-SCFI) Futura Rexolit™ or simil, Φ2″-4″ embedded in foundation slab and walls	20	u	SPV	PVC pipe Φ2″	20.40	20.00	m	3.24E+00

(continued)

Table 5 (continued)

Task		Task type						Material resources					
Code	Description	Code	Description	Specifications	Q	u		Code	Description	Deliver. quant.	Net quant.	Orig. unit	Normal quant. kg/m²
		F3G001	Gas piping	Copper pipe L-shaped (NMX-W-018-SCFI-2016) Nacobre™ or simil. Φ 1/2″ welded with tin/antimony (95/5) embedded in foundation slab and walls	14.50	m		MBC	Copper pipe	14.79	14.50	m	7.81E−02
G	Electric, lighting, communication systems	G1E001	Grid connection		1	u							2.02E−02
		G1E002	Main control table	3 100A keys Schneider Electric™	1	u							1.01E−02
		G1E003	Wiring	Wire calibre 12 y 14 (NOM-001-SEDE-2012) Condumex™ or simil. Corrugated plastic tube Φ 1/2″-3/4″ embedded in slab foundation and walls	35.60	m		MBC	Electric wire	36.67	35.60	m	1.44E−02
								SMP	Corrugated plastic tube	36.67	35.60	m	1.44E−01
		G1E004	Cieling socket	embedded resin octagonal box 4″ × 4″ Bticino™	5	u							1.37E−02
		G1E005	Socket	embedded resin box 2″ × 4″ for 3 modules Bticino™ for double contact with ivory plate Levinton™	11	u							8.20E−03

(continued)

Table 5 (continued)

| Task | | Task type | | | | | Material resources | | | | | |
Code	Description	Code	Description	Specifications	Q	u	Code	Description	Deliver. quant.	Net quant.	Orig. unit	Normal quant. kg/m²
		G1E006	Switches	Embedded resin box 2″ × 4″ for 3 modules Bücino™ for double contact with ivory plate Levinton™	6	u						8.20E−03
		G3T	Phone, fibre and tv coaxial cable sockets	Embedded resin box 2″ × 4″ for 3 modules Bücino™ for double contact with ivory plate Levinton™	5	u						1.64E−02
H	Windows, doors	H1A001	Al bedroom window	Anodized natural Al 1.20 × 1.50 XO lower fixed single pane 4 mm h: 0.30	3	u						8.38E−01
		H1A002	Al bathroom window	Anodized natural Al 1.20 × 1.50 XO lower fixed single pane 4 mm h: 0.30	2	u						2.31E−01
		H1A003	Al small window	Anodized natural Al 1.20 × 1.00 XO single pane 4 mm	1	u						2.15E−01
		H1A004	Al bathroom fixed window	Anodized natural Al 1.20 × 0.30 XO single pane 4 mm	1	u						1.15E−01
		H1A005	Kitchen door	White steel door Masonite™ 0.80 × 2.13	1	u						4.05E−01
		H2M001	Interior wooden door	Flush door (caobilla) 0.80 × 2.13 m transparent varnished, Al frame	2	u						8.28E−01
		H2M002	Bathroom door	Flush door (caobilla) 0.60 × 2.13 m transparent varnished, Al frame	1	u						3.10E−01
		H3A001	Main access door	Wooden frame, steel covered wooden door, 2 matte white panels, curve lintel single pane 3 mm	1	u						6.21E−01
Total												1.23E+03

Table 6 CDW transformation coefficients of material resources (CR)

Material resource	Waste origin	Produced waste	CR
Lime	Material loss	Lime	0.01
Lime	Paper	Paper/cardboard	1.00
Premixed concrete	Material loss	Concrete	0.07
Onsite made concrete	Material loss	Concrete	0.09
Onsite made concrete	Paper	Paper/cardboard	1.00
Concrete block vault	Material loss	Concrete	0.02
Concrete block vault	Wooden pallet	Wood	1.00
Concrete block vault	Plastic cover	Plastic	1.00
Concrete block 0.15 × 0.20 × 0.40 m	Material loss	Concrete	0.02
Concrete block 0.15 × 0.20 × 0.40 m	Wooden pallet	Wood	1.00
Concrete block 0.15 × 0.20 × 0.40 m	Plastic cover	Plastic	1.00
1:5 onsite made mortar	Material loss	Mixed	0.09
1:5 onsite made mortar	Paper	Paper/cardboard	1.00
Stucco	Material loss	Mixed	0.06
Stucco	Paper	Paper/cardboard	1.00
Cement-based adhesive	Material loss	Mixed	0.05
Cement-based adhesive	Paper	Paper/cardboard	1.00
Primary sealer	Material loss	Painting	0.05
Primary sealer	Tin	Steel/iron	1.00
Steel bar	Material loss	Steel/iron	0.03
Steel wire	Material loss	Steel/iron	0.01
Prefabricated steel column 0.12 × 0.18 × 3 m	Material loss	Steel/iron	0.08
Steel mesh	Material loss	Steel/iron	0.01
Steel mesh	Plastic cover	Plastic	1.00
Waterproof and insulating coating	Material loss	Painting	0.05
Waterproof and insulating coating	Metallic tin	Steel/iron	1.00
Polyester membrane roll	Material loss	Plastic	0.06
Polyester membrane roll	Plastic cover	Plastic	1.00
PVC tube Φ2″	Material loss	Plastic	0.02
Gypsum	Material loss	Gypsum	0.06
Gypsum	Paper	Paper/cardboard	1.00
Ceramic tile	Material loss	Ceramic	0.06
Ceramic tile	Cardboard	Paper/cardboard	1.00
Thermally fused PP water pipe	Material loss	Plastic	0.02
Plastic corrugated tube	Material loss	Plastic	0.03
Copper pipe 1/2″	Material loss	Copper	0.02

(continued)

Table 6 (continued)

Material resource	Waste origin	Produced waste	CR
Electric wire	Material loss	Copper	0.03
Electric wire	Cardboard	Paper/cardboard	1.00
Windows and doors	Plastic cover	Plastic	0.05

Table 7 Conventional construction model: CDW ton/m² and environmental indicators

CDW	Weight kg/m²	Energy kWh/m²	CO_{2e} emissions kg/m²	Reused CDW (RU) kg/m²	Onsite recycled CDW (RCo) kg/m²	Offsite recycled CDW (RCa) kg/m²	CDW to landfill (D) kg/m²
Inert soil	2.72E+01	7.55E−01	8.16E−01	0.00E+00	0.00E+00	2.72E+01	0.00E+00
Paper/cardboard	1.79E+00	1.62E−01	2.60E+00	0.00E+00	0.00E+00	2.63E−01	1.53E+00
Plastics	4.61E+00	9.60E−01	6.94E+01	0.00E+00	0.00E+00	2.30E−01	4.38E+00
Wooden pallet	6.46E+00	5.38E+00	4.07E+00	0.00E+00	0.00E+00	3.23E+00	3.23E+00
Ceramic	6.44E−01	2.33E−01	1.80E+00	0.00E+00	0.00E+00	1.29E+00	5.16E+00
Cement-based	6.92E+01	6.44E+00	1.84E+01	0.00E+00	0.00E+00	1.66E+00	6.63E+00
Mixed	1.74E−01	3.03E+00	4.41E−01	0.00E+00	0.00E+00	1.09E+01	4.36E+01
Steel/copper	5.47E−02	7.72E+00	1.81E−01	0.00E+00	0.00E+00	2.52E−01	3.94E−01
Others	8.29E−04	3.48E−01	1.37E−02	0.00E+00	0.00E+00	3.47E−02	1.99E−01
Total	1.10E+02	18.64E+01	9.77E+01	0.00E+00	0.00E+00	4.50E+01	6.51E+01

Table 8 Comparison between Mexican and Spanish CDW quantification models

Code Mexican model	Material resource	Weight ton/m²	Waste type ton/m²									Total
			Iron/steel	Paper/cardboard	Aggreg., cement and concrete	Ceramic	Wood	Plastics	Gypsum	Natural stones and terrazzo	Others	
CMC	Ceramic tile	1.09E−02		3.09E−04		6.46E−03						6.76E−03
F1H002	Shower set	6.07E−02		7.09E−06								7.09E−06
HAI	Adhesive Cemix	4.47E−04		3.26E−06								3.60E−06
HBC15	Concrete block 0.15 × 0.20 × 0.40 m	1.78E−01			5.71E−03		5.49E−03	3.84E−03				1.50E−02
HBO	Concrete block vault	2.16E−01			2.54E−03		9.71E−04					3.51E−03
HHO100	Onsite made concrete	4.96E−03		1.32E−06	4.51E−04							4.52E−04
HHO150	Onsite made concrete	3.10E−01		9.60E−05	2.92E−02							2.93E−02
HHO75	Onsite made concrete	1.71E−03		1.50E−07	1.56E−04							1.56E−04
HHP200	Premixed concrete	2.88E−01			2.13E−02							2.13E−02
HMM1-5	Onsite mortar 1:5	4.89E−02		2.38E−05	3.34E−03							3.36E−03
HMMP	Stucco Cemix™	1.74E−03		1.32E−03								1.32E−03
HRI	Primary sealer	4.71E−04						2.77E−05				2.77E−05
HVI	Prefab. reinforced concrete beam	7.20E−02										0.00E+00
MBC	Copper pipe	7.81E−05									2.64E−07	2.64E−07
MBC	Electric wire	1.44E−05		7.09E−06							6.49E−07	7.73E−06
NCL	Lime	1.01E−04		1.82E−07								1.82E−07
PBA	Insulating/waterproof painting × 200 l	1.68E−03	1.02E−04									1.02E−04

(continued)

Table 8 (continued)

Code Mexican model	Material resource	Weight ton/m²	Waste type ton/m²									Total
			Iron/steel	Paper/cardboard	Aggreg., cement and concrete	Ceramic	Wood	Plastics	Gypsum	Natural stones and terrazzo	Others	
PHV3-8	Steel bar	1.27E−02	5.06E−04									5.06E−04
PHVC-16	Steel wire	7.77E−05	7.00E−07									7.00E−07
PKC12	Prefab. steel framework ARMEX™ 0.12 × 0.18 m	4.23E−04	3.22E−05									3.22E−05
PME	Steel mesh	5.15E−03	3.09E−06						3.43E−04			3.46E−04
SMP	Thermally fused PP pipe	1.53E−02						3.64E−06				3.64E−06
SMP	Plastic corrugated tube	1.44E−04						4.32E−06				4.32E−06
SMS	Polyester membrane Thermotek™ × roll 109.25 m	4.20E−03						1.96E−04				1.96E−04
SPV	PVC pipe Φ2″	3.72E−03						2.69E−06				2.69E−06
W	Others	3.56E−03						1.90E−04			6.11E−05	2.51E−04
YYY	Gypsum	3.47E−03		2.17E−05					1.74E−04			1.95E−04
Mexican model		1.24E+00	6.44E−04	1.79E−03	6.27E−02	6.46E−03	6.46E−03	4.61E−03	1.74E−04		6.11E−05	8.30E−02
Spanish model		1.90E+00	6.20E−04	6.80E−04	6.80E−02	8.40E−03	4.40E−04	6.20E−04	6.80E−04	1.60E−04	3.80E−04	8.00E−02

References

Ajayi SO, Oyedele LO, Ceranic B, Gallanagh M, Kadiri KO (2015) Life cycle environmental performance of material specification: a BIM-enhanced comparative assessment. Int J Sustain Build Technol Urban Dev 6(1):14–24. https://doi.org/10.1080/2093761X.2015.1006708

Akinade OO, Oyedele LO, Ajayi SO, Bilal M, Alaka HA, Owolabi HA, Arawomo OO (2018) Designing out construction waste using BIM technology: stakeholders' expectations for industry deployment. J Clean Prod 180:375–385. https://doi.org/10.1016/j.jclepro.2018.01.022

Alderete Herrera JCA (2010) Vivienda de interés social. Universidad Veracruzana. Repositorio Institucional 5

Ambiente SM (n.d.) En marcha, Basura Cero en la Ciudad de México [WWW document]. Secretaría del Medio Ambiente. https://sedema.cdmx.gob.mx/comunicacion/nota/en-marcha-basura-cero-en-la-ciudad-de-mexico. Accessed 4.22.21

Araiza-Aguilar JA, Gutiérrez-Palacios C, Rojas-Valencia MN, Nájera-Aguilar HA, Gutiérrez-Hernández RF, Aguilar-Vera RA (2019) Selection of sites for the treatment and the final disposal of construction and demolition waste, using two approaches: an analysis for Mexico city. Sustainability 11:4077. https://doi.org/10.3390/su11154077

Argüello Méndez TR, Cuchí Burgos A (2008) Análisis del impacto ambiental asociado a los materiales de construcción empleados en las viviendas de bajo coste del programa 10×10 Con Techo-Chiapas del CYTED. Informes de la Construcción 60:25–34

Aslam MS, Huang B, Cui L (2020) Review of construction and demolition waste management in China and USA. J Environ Manag 264:110445. http://doi.org/10.1016/j.jenvman.2020.110445

Bakchan A, Faust KM, Leite F (2019) Seven-dimensional automated construction waste quantification and management framework: integration with project and site planning. Resour Conserv Recycl 146:462–474. http://doi.org/10.1016/j.resconrec.2019.02.020

Barón J, Conde J, Osuna M, Ramírez A, Solís J (2017) Consejería de Fomento y Vivienda/Vivienda y Rehabilitación/Base de Costes de la Construcción de Andalucía (BCCA), 29 abril 2016. Banco de precios

Blaisi NI (2019) Construction and demolition waste management in Saudi Arabia: current practice and roadmap for sustainable management. J Cleaner Prod 221:167–175. https://doi.org/10.1016/j.jclepro.2019.02.264

Cámara Mexicana de la Industria de la Construcción (2016) Plan de Manejo de Residuos de la Construcción y la Demolición

Carpio M, Roldán-Fontana J, Pacheco-Torres R, Ordóñez J (2016) Construction waste estimation depending on urban planning options in the design stage of residential buildings. Constr Build Mater 113: 561–570. https://doi.org/10.1016/j.conbuildmat.2016.03.061

CDMX (2019) Ahorrará Gobierno de la Ciudad de México para 2024, 8 mil 500 mdp con estrategia "Basura Cero" [WWW document]. CDMX. https://www.jefaturadegobierno.cdmx.gob.mx/com unicacion/nota/ahorrara-gobierno-de-la-ciudad-de-mexico-para-2024-8-mil-500-mdp-con-est rategia-basura-cero. Accessed 4.22.21

Cha G-W, Moon HJ, Kim Y-C, Hong W-H, Jeon G-Y, Yoon YR, Hwang C, Hwang J-H (2020) Evaluating recycling potential of demolition waste considering building structure types: a study in South Korea. J Cleaner Prod 256:120385. https://doi.org/10.1016/j.jclepro.2020.120385

Cheng JCP, Ma LYH (2013) A BIM-based system for demolition and renovation waste estimation and planning. Waste Manage 33:1539–1551. https://doi.org/10.1016/j.wasman.2013.01.001

Cochran KM, Townsend TG (2010) Estimating construction and demolition debris generation using a materials flow analysis approach. Waste Manage 30(11):2247–2254. https://doi.org/10.1016/j.wasman.2010.04.008

Comisión Económica para América Latina y el Caribe (2020) Contracción de la actividad económica de la región se profundiza a causa de la pandemia: caerá -9,1% en 2020 [WWW document]. https://www.cepal.org/es/comunicados/contraccion-la-actividad-eco nomica-la-region-se-profundiza-causa-la-pandemia-caera-91. Accessed 9.30.20

Concretos Reciclados [WWW document] (n.d.). http://www.concretosreciclados.com.mx/#3. Accessed 12.20.20

Consejo Nacional de Evaluación de la Política de Desarrollo Social (2018) Pobreza en México I CONEVAL [WWW document]. https://www.coneval.org.mx/Medicion/Paginas/PobrezaInicio. aspx. Accessed 9.28.20

Consejo Nacional de Evaluación de la Política de Desarrollo Social (CONEVAL) (2020) Informe de pobreza y evaluación 2020. Coahuila, México

Da Costa-Gómez O (2020) Buildings and construction [WWW document]. Mercado Interior, Industria, Emprendimiento y Pymes - European Commission. https://ec.europa.eu/growth/industry/sus tainability/built-environment_en. Accessed 1.13.21

de Guzmán Báez A, Villoria Sáez P, del Río Merino M, García Navarro J (2012) Methodology for quantification of waste generated in Spanish railway construction works. Waste Manage 32(5):920–924. https://doi.org/10.1016/j.wasman.2012.01.007

Ding T, Xiao J (2014) Estimation of building-related construction and demolition waste in Shanghai. Waste Manage 34(11):2327–2334. https://doi.org/10.1016/j.wasman.2014.07.029

European Commission (2008) Directive 2008/98/EC of the European Parliament and of the council of 19 November 2008 on waste and repealing certain directives

European Commission (2016) 2030 climate & energy framework [WWW document]. Climate action. European Commission. https://ec.europa.eu/clima/policies/strategies/2030_en. Accessed 2.9.21

European Commission (2018) EU construction and demolition waste protocol and guidelines

European Commission (2019) The European green deal

European Commission (2020a) Renovation wave [WWW document]. Energy. European Commission. https://ec.europa.eu/energy/topics/energy-efficiency/energy-efficient-buildi ngs/renovation-wave_en. Accessed 1.15.21

European Commission (2020b) A new circular economy action plan for a cleaner and more competitive Europe

European Commission (2020c) Level(s) the European framework for sustainable buildings [WWW document]. https://ec.europa.eu/environment/topics/circular-economy/levels_en. Accessed 1.15.21

European Environment Agency (2020) Construction and demolition waste: challenges and opportunities in a circular economy

Eurostat (2020a) Recovery rate of construction and demolition waste% of construction and demolition mineral waste recycled [WWW document]. Eurostat. https://ec.europa.eu/eurostat/ databrowser/view/CEI_WM040__custom_354944/bookmark/table?lang=en&bookmarkId=7e4 febc4-a0fd-444b-aae0-94643539ce0d. Accessed 1.18.21

Eurostat (2020b) Generation of waste by economic activity [WWW document]. Generation of waste by economic activity. https://ec.europa.eu/eurostat/databrowser/view/ten00106/default/table?lan g=en. Accessed 1.18.21

Ge XJ, Livesey P, Wang J, Huang S, He X, Zhang C (2017) Deconstruction waste management through 3d reconstruction and bim: a case study. Vis Eng 5:13. https://doi.org/10.1186/s40327-017-0050-5

Ghosh A, Nundy S, Ghosh S, Mallick TK (2020) Study of COVID-19 pandemic in London (UK) from urban context. Cities 106:102928. http://doi.org/10.1016/j.cities.2020.102928

Gobierno de España (2020) Estrategia Española de Economía Circular ESPAÑA CIRCULAR 2030

Gobierno de México (2015) Informe de la Situación del Medio Ambiente en México

Gobierno Federal de México (2016) Código de Edificación de Vivienda 2010

GODF (2015) Norma Ambiental para el Distrito Federal NADF-007-RNAT-2013, que establece la clasificación y especificaciones de manejo para residuos de la construcción y demolición, en el Distrito Federal; actualizada el 26 de febrero de 2015. Gaceta Oficial del Distrito Federal

Guerra BC, Bakchan A, Leite F, Faust KM (2019) BIM-based automated construction waste estimation algorithms: the case of concrete and drywall waste streams. Waste Manage 87:825–832. https://doi.org/10.1016/j.wasman.2019.03.010

Hendriks CF (ed) (2000) Sustainable raw materials: construction and demolition waste. RILEM report. RILEM Publ, Cachan

Hyman M, Turner B, Carpintero A, United Nations Institute for Training and Research, Inter-Organization Programme for the Sound Management of Chemicals, United Nations Environment Programme (2015) Guidelines for national waste management strategies: moving from challenges to opportunities

ITEC Instituto de la construcción de Catalunia (n.d.) Información ambiental de productos y sistemas [WWW document]. https://itec.es/metabase/productos-sostenibles/0/0/0/. Accessed 2.9.21

Jaillon L, Poon CS, Chiang YH (2009) Quantifying the waste reduction potential of using prefabrication in building construction in Hong Kong. Waste Manage 29(1):309–320. https://doi.org/10.1016/j.wasman.2008.02.015

Jalaei F, Zoghi M, Khoshand A (2019) Life cycle environmental impact assessment to manage and optimize construction waste using Building Information Modeling (BIM). Int J Constr Manag 1–18. http://doi.org/10.1080/15623599.2019.1583850

Jefatura de Gobierno (2011) Ley 22/2011 Residuos y Suelos Contaminados

Jiménez Rivero A, Sathre R, García Navarro J (2016) Life cycle energy and material flow implications of gypsum plasterboard recycling in the European Union. Resour Conserv Recycl 108:171–181. https://doi.org/10.1016/j.resconrec.2016.01.014

Jin R, Yuan H, Chen Q (2019) Science mapping approach to assisting the review of construction and demolition waste management research published between 2009 and 2018. Resour Conserv Recycl 140:175–188. https://doi.org/10.1016/j.resconrec.2018.09.029

Kabirifar K, Mojtahedi M, Wang C, Tam VWY (2020) Construction and demolition waste management contributing factors coupled with reduce, reuse, and recycle strategies for effective waste management: a review. J Cleaner Prod 263:121265. http://doi.org/10.1016/j.jclepro.2020.121265

Kartam N, Al-Mutairi N, Al-Ghusain I, Al-Humoud J (2004) Environmental management of construction and demolition waste in Kuwait. Waste Manage 24(10):1049–1059. https://doi.org/10.1016/j.wasman.2004.06.003

Kleemann F, Lehner H, Szczypińska A, Lederer J, Fellner J (2017) Using change detection data to assess amount and composition of demolition waste from buildings in Vienna. Resour Conserv Recycl 123:37–46. https://doi.org/10.1016/j.resconrec.2016.06.010

Klevnäs P, Kulldorff A (2020) The circular economy and covid-19 recovery how pursuing a circular future for Europe fits with recovery from the economic crisis. Material Economics

Li J, Ding Z, Mi X, Wang J (2013) A model for estimating construction waste generation index for building project in China. Resour, Conserv Recycl 74:20–26. https://doi.org/10.1016/j.resconrec.2013.02.015

Li CZ, Zhao Y, Xiao B, Yu B, Tam VWY, Chen Z, Ya Y (2020) Research trend of the application of information technologies in construction and demolition waste management. J Cleaner Prod 263:121458. http://doi.org/10.1016/j.jclepro.2020.121458

Liu Z, Osmani M, Demian P, Baldwin A (2015) A BIM-aided construction waste minimisation framework. Autom Constr 59:1–23. https://doi.org/10.1016/j.autcon.2015.07.020

Liu H, Sydora C, Altaf MS, Han S, Al-Hussein M (2019) Towards sustainable construction: BIM-enabled design and planning of roof sheathing installation for prefabricated buildings. J Clean Prod 235:1189–1201. https://doi.org/10.1016/j.jclepro.2019.07.055

López-Feldman A (2014) Cambio climático, distribución del ingreso y la pobreza. El caso de México, Estudios del cambio climático en América Latina. Comisión Económica para América Latina y el Caribe (CEPAL), Santiago de Chile

López-López J (2019) Caracterización de los residuos de la construcción de la vivienda en México. Un modelo teórico. Universidad de Sevilla, Sevilla

López-Mesa B, Pitarch Á, Tomás A, Gallego T (2009) Comparison of environmental impacts of building structures with in situ cast floors and with precast concrete floors. Build Environ 44:699–712. https://doi.org/10.1016/j.buildenv.2008.05.017

Lu W, Chen X, Peng Y, Shen L (2015) Benchmarking construction waste management performance using big data. Resour, Conserv Recycl 105:49–58. https://doi.org/10.1016/j.resconrec. 2015.10.013

Luciano A, Reale P, Cutaia L, Carletti R, Pentassuglia R, Elmo G, Mancini G (2018) Resources optimization and sustainable waste management in construction chain in Italy: toward a resource efficiency plan. Waste Biomass Valor. https://doi.org/10.1007/s12649-018-0533-1

Luciano A, Cutaia L, Cioffi F, Sinibaldi C (2020) Demolition and construction recycling unified management: the DECORUM platform for improvement of resource efficiency in the construction sector. Environ Sci Pollut Res. https://doi.org/10.1007/s11356-020-09513-6

Matthew RA, McDonald B (2006) Cities under siege: urban planning and the threat of infectious disease. J Am Plann Assoc 72:109–117. https://doi.org/10.1080/01944360608976728

Medina Ross JA, México, Secretaría de Medio Ambiente y Recursos Naturales (2001) Minimización y manejo ambiental de los residuos sólidos. Secretaría de Medio Ambiente y Recursos Naturales, México

Menegaki M, Damigos D (2018) A review on current situation and challenges of construction and demolition waste management. Curr Opin Green Sustain Chem. Reuse and recycling/UN SGDs: how can sustainable chemistry contribute? Green Chem Educ 13:8–15. http://doi.org/10.1016/j. cogsc.2018.02.010

Mercader Moyano M (2010) Cuantificación de los recursos consumidos y emisiones de CO_2 producidas en las construcciones de Andalucía y sus implicaciones en el protocolo de Kioto (tesis doctoral). Universidad de Sevilla, Sevilla

Mercader-Moyano P, Ramírez-de-Arellano-Agudo A (2013) Selective classification and quantification model of C&D waste from material resources consumed in residential building construction. Waste Manag Res 31:458–474. https://doi.org/10.1177/0734242X13477719

Mercader Moyano MP, Camporeale PE, Cózar-Cózar E (2019) Evaluación de impacto ambiental mediante la introducción de indicadores a un modelo BIM de vivienda social. HS 9:78–93. http://doi.org/10.22320/07190700.2019.09.02.07

Miatto A, Schandl H, Forlin L, Ronzani F, Borin P, Giordano A, Tanikawa H (2019) A spatial analysis of material stock accumulation and demolition waste potential of buildings: a case study of Padua. Resour Conserv Recycl 142:245–256. https://doi.org/10.1016/j.resconrec.2018.12.011

Ministerio de Agricultura, Alimentación y Medio Ambiente (2015a) Real Decreto 180/2015 Traslado de residuos en el interior del territorio del Estado

Ministerio de Agricultura, Alimentación y Medio Ambiente (2015b) Plan Estatal Marco de Gestión de Residuos (PEMAR) 2016–2022

Ministerio de la Presidencia (2008) Real Decreto 105/2008. La producción y gestión de los residuos de construcción y demolición

Molar-Orozco ME, Velázquez-Lozano J, Vázquez-Jimánez MG (2020) Comportamiento térmico de tres prototipos en Saltillo, Coahuila (bloques de tierra, concreto y tapa de huevo). HS 10:22–31. http://doi.org/10.22320/07190700.2020.10.01.02

OECD (2020) The global economy risks falling ill. Interim report, March 2020 [WWW document]. OECD Economic Outlook. http://www.oecd.org/economic-outlook/. Accessed 1.13.21

Official Journal of the European Communities. Commission Decision 2001/118/EC of wastes, 2001. European waste catalogue

Pacheco-Torgal F (2014) Eco-efficient construction and building materials research under the EU framework programme horizon 2020. Constr Build Mater 51:151–162. https://doi.org/10.1016/j.conbuildmat.2013.10.058

Park JW, Cha GW, Hong WH, Seo HC (2014) A study on the establishment of demolition waste DB system by BIM-based building materials [WWW document]. Appl Mech Mater. https://doi.org/10.4028/www.scientific.net/AMM.522-524.806

Parisi Kern A, Ferreira Dias M, Piva Kulakowski M, Paulo Gomes L (2015) Waste generated in high-rise buildings construction: a quantification model based on statistical multiple regression. Waste Manage 39: 35–44. https://doi.org/10.1016/j.wasman.2015.01.043

Procuraduría Federal de Protección al Ambiente (2014) Norma Oficial Mexicana NOM-161-SEMARNAT-2011

Rakhshan K, Morel J-C, Alaka H, Charef R (2020) Components reuse in the building sector—a systematic review. Waste Manag Res 38:347–370. https://doi.org/10.1177/0734242X20910463

Ram V, Kalidindi SN (2017) Estimation of construction and demolition waste using waste generation rates in Chennai, India: waste Management & Research. J Sustain Circular Econ 35(6):610–617. https://doi.org/10.1177/0734242X17693297

Ramírez de Arellano Agudo A (2002) Retirada selectiva de residuos: Modelo de presupuestación. Colegio Oficial de Aparejadores y Arquitectos Técnicos de Sevilla, Sevilla

Ramírez de Arellano Agudo A (2014) Presupuestación de obras. Secretariado de Publicaciones, Universidad de Sevilla, Sevilla

Ríos O (2019) Instituto de ecología [WWW document]. Instituto de Ecología UNAM. http://www.ecologia.unam.mx/web/index.php?option=com_content&view=article&id=193. Accessed 4.22.21

Roux Gutierrez RS, García Izaguirre VM, Espuna Mujica JA (2015) Los materiales alternativos estabilizados y su impacto ambiental. Nova Scientia 7:243–266

Sánchez-Corral J (2013) La vivienda "social" en México pasado, presente, futuro? Sistema Nacional de Creadores de Arte- JSE, México

Secretaría del medio ambiente (2018) Norma ambiental para el D.F. NADF-024-AMBT-2013, que establece los criterios y especificaciones técnicas bajo las cuales se deberá realizar la separación, clasificación, recolección selectiva y almacenamiento de los residuos del Distrito Federal

Secretaría de Medio Ambiente y Recursos Naturales (2010) Estudio de análisis, evaluación y definición de estrategias de solución de la corriente de residuos generada por las actividades de construcción en México

Tanikawa H, Hashimoto S (2009) Urban stock over time: spatial material stock analysis using 4d-GIS. Build Res Inf 37:483–502. https://doi.org/10.1080/09613210903169394

Turcott Cervantes DE, Romero EO, del Consuelo Hernández Berriel M, Martínez AL, del Consuelo Mañón Salas M, Lobo A (2021) Assessment of some governance aspects in waste management systems: a case study in Mexican municipalities. J Cleaner Prod 278:123320. http://doi.org/10.1016/j.jclepro.2020.123320

United Nations Environment Programme (2018) Waste management outlook for Latin America and the Caribbean

Villoria Sáez P, Santa Cruz Astorqui J, del Río Merino M, Mercader Moyano MP, Rodríguez Sánchez A (2018) Estimation of construction and demolition waste in building energy efficiency retrofitting works of the vertical envelope. J Cleaner Prod 172:2978–2985. https://doi.org/10.1016/j.jclepro.2017.11.113

Won J, Cheng JCP (2017) Identifying potential opportunities of building information modeling for construction and demolition waste management and minimization. Autom Constr 79:3–18. https://doi.org/10.1016/j.autcon.2017.02.002

World Bank (2020) Global economic prospects, June 2020. World Bank, Washington, DC. https://doi.org/10.1596/978-1-4648-1553-9

Wu Z, Yu ATW, Shen L, Liu G (2014) Quantifying construction and demolition waste: an analytical review. Waste Manage 34:1683–1692. https://doi.org/10.1016/j.wasman.2014.05.010

Wu H, Duan H, Zheng L, Wang J, Niu Y, Zhang G (2016) Demolition waste generation and recycling potentials in a rapidly developing flagship megacity of South China: prospective scenarios and implications. Constr Build Mater 113:1007–1016. https://doi.org/10.1016/j.conbuildmat.2016.03.130

Wu Z, Yu ATW, Poon CS (2019) An off-site snapshot methodology for estimating building construction waste composition—a case study of Hong Kong. Environ Impact Assess Rev 77:128–135. https://doi.org/10.1016/j.eiar.2019.03.006

Wu H, Zuo J, Zillante G, Wang J, Yuan H (2019) Status quo and future directions of construction and demolition waste research: a critical review. J Cleaner Prod 240:118163. https://doi.org/10.1016/j.jclepro.2019.118163

Xu C, Jia M, Xu M, Long Y, Jia H (2019) Progress on environmental and economic evaluation of low-impact development type of best management practices through a life cycle perspective. J Clean Prod 213:1103–1114. https://doi.org/10.1016/j.jclepro.2018.12.272

Yajnes M, Caruso S, Kozak D, Kozak A, Mühlmann S (2017) Gestión de residuos y producción de bloques con material reciclado in situ en una obra de escala intermedia en la ciudad de Buenos Aires, Argentina, in: 3er Congreso Internacional de Construcción Sostenible y Soluciones Eco-Eficientes. 3er Congreso Internacional de Construcción Sostenible y Soluciones Eco-Eficientes, Universidad de Sevilla, Departamento de Construcciones Arquitectónicas I, Sevilla, pp 1035–1059

Yuan H (2017) Barriers and countermeasures for managing construction and demolition waste: a case of Shenzhen in China. J Cleaner Prod 157:84–93. https://doi.org/10.1016/j.jclepro.2017.04.137

Zahradnik P, Palmieri S, Dirx J (2020) Resolución sobre las propuestas del CESE para la reconstrucción y la recuperación tras la crisis de la COVID-19

Zaman AU, Lehmann S (2013) The zero waste index: a performance measurement tool for waste management systems in a 'zero waste city'. J Cleaner Prod 50:123–132. https://doi.org/10.1016/j.jclepro.2012.11.041

Ecological Footprint Assessment of Recycled Asphalt Pavement Construction

Ansari Yakub Zafar Abid, Ansari Abu Usama, Dilawar Husain, Manish Sharma, and Ravi Prakash

Abstract The replacement of bitumen with waste/recycled plastics in asphalt pavement is an effective way to achieve sustainability goals in road construction. The use of waste/recycled plastics in asphalt pavements requires an emphasis on environmental assessment of such asphalt pavements. The study develops a methodology to assess the Ecological Footprint (EF) of asphalt pavement construction. It also evaluates the proposed Sustainable Recycling Index (SRI) of different asphalt pavements based on normalized ecological footprint and cost parameters. In this case study, waste plastic replaced bitumen by the ratios of 10% (w/w), 20% (w/w), 30% (w/w) and 40% (w/w) in the asphalt pavement. The EF of conventional asphalt pavement construction (10 m wide and 1 km with 200 mm thickness) is about 61.78 gha. The EF reduction potential of asphalt pavement construction is 4.1%, 8.2%, 12.3% and 16.4% for 10%, 20%, 30% and 40% of waste plastic used in place of total bitumen in the asphalt pavement, respectively. The SRI of asphalt pavement with 10%, 20%, 30% and 40% waste plastic are about 0.002, 0.007, 0.016 and 0.029, respectively. Hence the replacement of bitumen with waste plastic in asphalt pavement gives significant results, demonstrating that the use of waste plastic in asphalt pavement is ecofriendly and economical.

Keywords Ecological footprint · Asphalt pavement · Sustainability · Recycling · Waste management

A. Y. Z. Abid · A. A. Usama
Department of Civil Engineering, Maulana Mukhtar Ahmad Nadvi Technical Campus, Malegaon, Maharashtra 423203, India

D. Husain (✉)
Department of Mechanical Engineering, Maulana Mukhtar Ahmad Nadvi Technical Campus, Malegaon, Maharashtra 423203, India

M. Sharma
Department of Mechanical Engineering, Malla Reddy Engineering College, Hyderabad 500100, India

R. Prakash
Department of Mechanical Engineering, Motilal Nehru National Institute of Technology, Allahabad 211004, India

© The Author(s), under exclusive license to Springer Nature Singapore Pte Ltd. 2022 137
S. S. Muthu (ed.), *Environmental Footprints of Recycled Products*,
Environmental Footprints and Eco-design of Products and Processes,
https://doi.org/10.1007/978-981-16-8426-5_5

1   Introduction

India annually generates 5.6 million tonnes of plastic waste according to the Central Pollution Control Board (CPCB) report. Waste plastic or plastic pollution involves the growth of plastic goods in the eco-system that badly affects wildlife, marine and human lifestyle. In general, plastics pollutants are categorized as macro-debris, based on size. The status of plastic pollution is associated with plastics being cheap and robust, it promotes the use of plastics for different purposes. However, degradation rate of plastic is very slow. Waste plastic adversely affects ground, waterways and oceans, etc. Organisms, like animals and predominantly marine animals, are also affected through direct assimilation of plastic waste, or through exposure to chemicals within plastics that cause interruptions in biological functions. Thus, it's very necessary to solve the problem of waste plastic. The potential use of plastics in the construction of asphalt pavement reduces the amount of waste plastic. Presently polypropylene, polyester, mineral and cellulose are commonly used as fibres. In this present study, polyethylene is used as stabilizing additive to improve performance characteristics of pavement.

According to Federal Highway Administration (FHWA) recycled materials policy, the three mandatory requirements that must be incorporated into asphalt pavement recycling are as follows (FHWA 2021):

1. Economically feasible
2. Ecofriendly nature
3. Enhance performance.

Asphalt Pavement

The asphalt pavement consists of aggregate, sand and bitumen. Bitumen has behaved as binding material in asphalt pavements. It is viscous and semi-solid form at NTP conditions. In general, the pavement composition is 90–95% aggregate and sand and 5–10% bitumen. The pavement is better known for its durability, resilience and flexibility. The asphalt pavement can easily change the shape according to the base shape change that occurs due to weather condition. It makes asphalt pavement leading option for road construction. A typical cross-section view of asphalt pavement layers is depicted in Fig. 1.

The surface wearing coarse is the top layer of asphalt pavement (thickness of 2–4 in.), it comes directly in contact with traffic loads. It is generally constructed with superior quality/dense grade asphalt concrete and construct on the base course layer of pavement. It provides friction as well as smoothness for better driving and prevent wear and tear in wheels. It also provides water proofing to avoid weakening effects in base and subbase layers due to water (Mannering and Washburn 2013). The base course layer (thickness 4–10 in.) and subbase course layer (thickness 4–10 in.) consist of different sizing aggregates. The base layer is placed on sub base layer to support surface wearing course. The subbase layer is placed on subgrade layer; it helps to support base course. Waste materials like construction and demolition

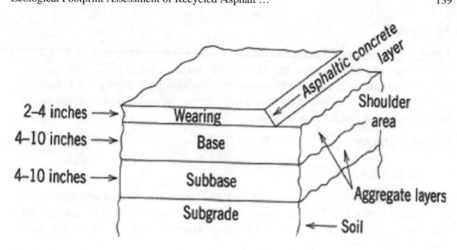

Fig. 1 Typical cross-section views of asphalt pavement layers (Mannering and Washburn 2013)

waste, municipal waste iron slag, etc. can use to construct sub base course layer. The subgrade layer consists of compacted soil and performs as a foundation for the pavement.

Asphalt Pavement Materials

The details of asphalt pavement materials are listed below:

Bitumen or Bituminous

Bitumen is viscous, semi-solid liquid and black or brown in colour. It contains long-chain hydrocarbons. It is produced during the petroleum/fossil fuel refinement and it is also found as a constituent of naturally occurring asphalt. It is significantly non-volatile and easily become soft during heating. It has adhesive property as well as behave like waterproofing material.

Classification of Bitumen/Bituminous

The general classification of bitumen or bituminous is as follows:

Bitumen 80/100	"The characteristics of this grade confirm to that of S 90 grade of IS-73-1992. This is the softest of all grades available in India. This is suitable for low volume roads and is still widely used in the country"
Bitumen 60/70	"This grade is harder than 80/100 and can withstand higher traffic loads. The characteristics of this grade confirm that of S 65 grade of IS-73-1992. It is presently used mainly in construction of National Highways and State Highways"
Bitumen 30/40	"This is the hardest of all the grades and can withstand very heavy traffic loads. The characteristics of this grade confirm to that of S 35 grade of IS-73-1992"
Bitumen 30/40	"It is used in specialized applications like airport runways and also in very heavy traffic volume roads in coastal cities in the country"

Aggregates

Aggregates contribute the highest share in pavement construction. It is the most voluminous ingredient and it has the highest mixture weight (90–95%) in pavement construction. In general, pavement construction used aggregates obtained by crushed rocks. Its categorized into three main groups of rocks: first, Igneous, second, Sedimentary and third, Metamorphic.

Type of aggregates

(a) Coarse aggregates

The coarse aggregates are retained on 4.75 mm IS sieve size. Coarse aggregate should be screened crushed rock, angular in shape, free from organic matters, dust particles and clay, etc. It gives good compressive strength and shear strength as well as shows interlocking properties. Fine aggregates.

The fine aggregate is retained on 0.075 mm IS sieve. It should be clean screened dust and should be free from organic matter, loam and clay. It consists of stone crusher dust. The details of size of aggregate are mentioned in Table 1.

Asphalt Pavement Thickness

Indian asphalt pavements are designed as per the IRC:37-2018 code, the thickness of the pavement depends on two parameters: (1) traffic load and (2) California Bearing Ratio (CBR). The pavement thickness designs given in the IRC:37-2018 are applicable to design traffic up to 50 million standard axles (msa) and CBR up to 15% (IRC 2018). The asphalt pavement thickness of different sections is depicted in Fig. 2. According to the IRC:37-2018, the asphalt pavement thickness increases with the traffic load while it decreases with increase of CBR value. The significant changes come in Surface Course and Base Course as compared to rest of the pavement sections for different traffic loads and CBR values. The asphalt concrete thickness varies from 80 to 180 mm for the different traffic loads and CBR values.

Asphalt Pavement Construction Process

The asphalt production is directly associated with the petroleum/crude oil refining sector.

Asphalt production includes different processes: (1) Bitumen production, (2) Aggregate extraction (3) Asphalt storage (4) Asphalt mixing and aggregate drying (5) Pavement placement. Energy consumption and GHG emissions for asphalt production (includes crude oil extraction, transport and refining, etc.) are 4900 MJ/t and

Table 1 Details of aggregate size

Types of aggregates	Size (mm)
Coarse aggregate	4.75–40 mm
Fine aggregate	0.075–4.75 mm
Filler or dust	<0.075 mm

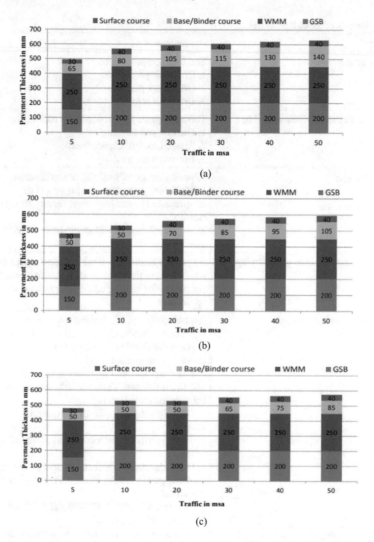

Fig. 2 Pavement thickness in mm versus traffic in msa **a** for CBR 5%, **b** for CBR 10%, **c** for CBR 10% (IRC 2018)

285 kgCO$_2$/t, respectively (Chehovits and Galehouse 2010). It indicates that asphalt production is a high energy-intensity and emission process. The asphalt production details are mentioned in Table 2.

Recycling Materials in Asphalt Pavement

Various recycled materials are widely used in asphalt pavement (i.e. construction and demolition waste, reclaimed asphalt pavement, waste rocks, cement dust, glass, crumb rubber, steel slag, waste plastic, waste engine oils, etc.) to minimize the

Table 2 Details of asphalt production

Processes	Production
Bitumen production	The bitumen production process: 1. Crude oil production 2. Oil transportation 3. Crude oil refining
Aggregates extraction (coarse and fine aggregate)	The aggregates production from rock mass: 1. Drilling and blasting 2. Blasted rock transportation 3. Blasted rock crushing
Asphalt storage	Asphalt storage facilities have significance: 1. Maintaining the asphalt at appropriate temperature is most important for any storage facility 2. In general, hot oil and steam both are used for heating asphalt
Asphalt mixing and aggregates drying	Aggregate heating (drying) is necessary for asphalt pavement construction. Large amount of energy is consumed for heating and drying process of aggregates
Asphalt pavement placement	Asphalt pavement placement required different types of machines. Use of machines directly impacts the GHG emissions due to fossil fuel consumption in machines For any asphalt plant production capacity of 360 t/h, machineries and fuel (diesel) consumption for pavement placement: 1. One asphalt paver (diesel consumption rate 15.1 l/h) 2. Three rollers (diesel consumption rate 17.0 l/h each) 3. One tack truck (diesel consumption rate 26.5 l/h) 4. Three pickup trucks (diesel consumption rate 4.2 l/h each) 5. One small loader (CAT 950 sized) (diesel consumption rate 26.5 l/h) 6. One small broom (diesel consumption rate 17.0 l/h)

demands of conventional/natural recourses as well as reduce the environmental impact of the pavement (Austoroads 2009). It also helps to manage millions of cubic meters of waste materials that are mostly disposed as landfills.

Recycled materials also enhance the toughness, durability, plasticity and stiffness of pavement construction. The economic benefits can also achieve by using low-cost recycled materials, which reduce the overall cost of pavement construction (Reid 1998). Recycled materials have the potential to fulfil all criteria of sustainable pavement construction. Sustainability goals can be achieved by efficient use of recycled materials in pavement construction because natural materials demand, GHG emissions and discharge of contaminants into the land are diminished.

Role of Waste Plastics in Asphalt Pavements

Now a day, waste plastic is generally used in asphalt pavement construction for enhancing characteristic properties like durability, smoothness, strength, etc. (Al-Hadidy and Yi-qiu 2009). Asphalt concrete becomes weak when it comes to contact with water. However, the addition of waste plastics into asphalt concrete led to the enhancement of the strength, water repellent properties of asphalt concrete. Waste plastics (i.e. water bottles, disposable cups, polyethylene packets and plastic carry-bags, etc.) can be collected from garbage and used as an ingredient of the pavement construction. These packing materials are made up of Low-density polyethylene (LDPE), Polystyrene (PS), Polyethylene (PE), Polypropylene (PP), Biaxially Oriented Polypropylene (BOPP) and Polyethylene terephthalate (PET), etc. The waste plastics products are characterized by the thickness of the materials and softening temperature of the materials. The details of the waste plastic products are depicted in Table 3.

Waste plastics have a good adhesion property, it increases the compression strength and bending strength of pavement structure. Therefore, waste plastics can be used as a binder in asphalt pavement. The pavement strength increases with the increase of plastic content in asphalt pavement. It is depended on the types of plastics used in pavement. The increasing order of pavement strength with the use of different plastics is PS < PE < PP < Laminated films < BOPP.

Table 3 The details of the waste plastic products (Vansudevan et al. 2012)

Commercial plastic material	Nature of plastics	Softening point (°C)	Thickness (μ)
Water bottle	PET	170–180	210
Carry bag	PE	100–120	10
Milk pouch	LDPE	100–120	60
Chocolate covers	Polyester + PE + metalized polyester	155	20
Cold drinks bottle	PET	170–180	210
Foam	PE	100–110	NA
Parcel cover	PE	100–120	50
Cup	PE	100–120	150
Biscuit covers	Polyester + PE	170	40
Film	PE	120–130	50
Decoration papers	BOPP	110	100
Foam	PS	110	NA

Waste plastic can be added directly to hot asphalt and the asphalt concrete with waste plastic is placed on the road surface like a normal asphalt pavement. The modified asphalt mixture by using waste plastics (e.g. waste plastic) which enhances the properties of Hot Mix Asphalt mixtures would produce a more durable pavement. It helps recycle waste plastics and also protect environment by managing waste disposal of waste plastics. The steps involve in waste plastics mixing in asphalt pavement is depicted in Fig. 3.

Ecological Footprint (EF): The EF assessment has been developed by Mathis Wackernagel and William Rees (Wackernagel and Rees 1996). The EF assessment includes all resources as input parameters and transforms them into a single output (global hectare). The EF indicator can be utilized to estimate the various types of sustainable measures such as the feasibility of appropriate distribution of resources of the planet. The global hectare (gha) is defined as "one gha is equivalent to one hectare of bio-productive land with world average productivity".

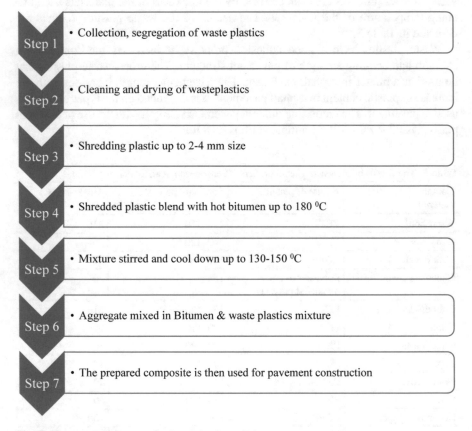

Fig. 3 The process of waste plastic mixing in asphalt pavement

2 Methodology

The bio-productive lands are significant factors for asphalt pavement. A method has been developed to examine the manufacturing and placement of different waste plastics compositions in asphalt pavement through the EF assessment. The details of EF assessment of recycled pavements are as follows:

2.1 Ecological Footprint of Asphalt Pavement (EF$_P$)

The EF$_P$ of asphalt pavement has been assessed in this study. The EF$_P$ assessment of asphalt pavement includes two components: first, raw materials and manufactured pavement materials and second, placement of pavement. In this study, the transportation of pavement materials is not considered. The system boundary for the EF assessment of asphalt pavement is shown in Fig. 4.

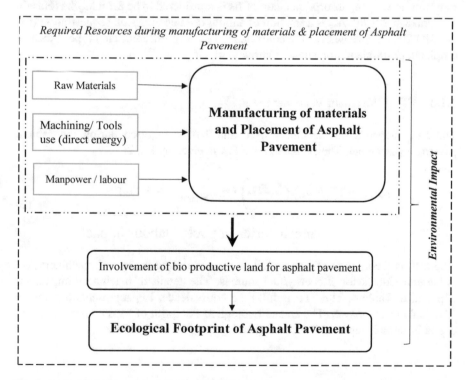

Fig. 4 System boundary of asphalt pavement construction

2.1.1 EF of Raw and Manufactured Materials (EF$_{pm}$)

Raw and Manufactured materials of asphalt pavement are accountable for substantial resource consumption therefore, the ecological impact of asphalt pavement should be examined. The EF of raw and manufactured pavement materials depends on three factors: (1) material consumptions, (2) direct energy (machinery) uses during materials manufacturing. The physical land use during material extraction is neglected in this case study. The EF$_{pm}$ has been calculated by Eq. 1 (Husain and Prakash 2019a):

$$EF_{pm} = \underbrace{\sum \left(\frac{C_{mi} \cdot E_{mi}}{A_f/(1-A_{oc})}\right) \cdot e_{CO_2 \text{ land}}}_{\text{manufacturing impact}} + \underbrace{\sum \left(\frac{C_{wi}}{Y_{wi}}\right) \cdot e_i}_{\text{natural impact}} \tag{1}$$

where, C_{mi} has represented material consumption of ith material, E_{mi} has embodied emission of ith material. C_{wi} is consumption in ith natural material and Y_{wi} is materials productivity. A_f absorption factor of fts is considered to be 2.7 tCO$_2$/ha (Husain and Prakash 2019b), A_{oc} is a fraction of annual oceanic emission sequestration (i.e. 0.3 SIO 2017). e_i represents the equivalence factor of different land types (such as cropland, pastureland, forestland, marine land, etc.).

2.1.2 EF of Pavement Placement (EF$_{pp}$)

The EF$_{pp}$ depends on machinery used and labour required during the asphalt pavement placement. The estimation of EF$_{pp}$ is given by Eq. 2:

$$EF_{pp} = \underbrace{\sum (E_i \cdot \alpha_i) \cdot \left(\frac{1-A_{oc}}{A_f}\right) \cdot e_{CO_2 \text{ land}}}_{\text{machineries impact}} + \underbrace{FTE \cdot EF_l}_{\text{labour impact}} \tag{2}$$

where E_i is the amount of energy/fuel consumed during the use of machinery; α_i. is the emission factor of energy/fuel sources. The details of machinery impact are depicted in Table 4. The FTE is full-time equivalence, EF$_l$ represents the annual EF of labour/manpower. The annual Ecological Footprint of food consumption per capita in India is reported in Table 5 (Husain and Prakash 2019c).

Table 4 The details of EF of machinery

Machineries	Fuel type	Consumption rate (l/h)	Emissions rate (kgCO$_2$/kg of fuel) (EEA 2013)	Emissions rate (kgCO$_2$/l of fuel)	EF (gha/h)
Asphalt paver	Diesel	15.1	3.17	3.73	0.013502
Rollers	Diesel	17.0	3.17	3.73	0.015201
Tack truck	Diesel	26.5	3.17	3.73	0.023696
Pickup trucks	Diesel	4.2	3.17	3.73	0.003756
Small loader (CAT 950 sized)	Diesel	26.5	3.17	3.73	0.023696
Small broom	Diesel	17.0	3.17	3.73	0.015201

Table 5 Annual ecological footprint of food consumption per capita in India

	Monthly consumption (NSSO 2014)	Total annual consumption	CO$_2$ emission factor	Yield production (tonne/ha)	EF (gha)
Cereals	9.28 kg	111.36 kg		2.39 (OGD 2019)	0.117
Pulses	0.90 kg	10.8 kg		0.69 (OGD 2019)	0.039
Vegetable	8.4 kg	100.8 kg		1.61 (OGD 2019)	0.157
Beef	0.06 kg	0.72 kg		32 (Chambers et al. 2004)	0.011
Mutton	0.08 kg	0.96 kg		72	0.006
Milk	5.4 l	64.8 l		458	0.062
Fish	0.252 kg	3.024 kg		0.035	0.030
Fruits	0.654 kg	7.848 kg		2330	1.58×10^{-6}
Edible oil	0.85 kg	10.2 kg		0.38	0.068
Wood	4.3 kg	51.6 kg	1.5–1.6 (kgCO$_2$/kg)	73 m^3/ha (IHD 2014)	0.028
LPG	1.9 kg	22.8 kg	3.31 (kgCO$_2$/kg)		0.025
Kerosene	0.40 l	4.8 l	2.58 (kgCO$_2$/l)		0.004
Total annual EF/person					**0.549**

2.2 Sustainable Recycling Index (SRI)

Sustainability is a comprehensive concept that contains human activities and their influences on the planet. The product-based indices are developed for comparison of similar kinds of products as well as they suggest which one is more sustainable in a certain boundary. To measure and compare the sustainability of pavement recycling, the Sustainability Recycling Index (SRI) has been developed based on two parameters: (1) Ecological Footprint of pavement and (2) Cost of pavement. The SRI is a simple and effective tool to compare pavement recycling, it also suggests the limitations in pavement recycling. The expression of the SRI is as follows:

$$\text{SRI} = \frac{\text{EF}_{\text{Base}} - \text{EF}_{\text{Recycle}}}{\text{EF}_{\text{Base}}} \times \frac{C_{\text{Base}} - C_{\text{Recycle}}}{C_{\text{Base}}}.$$

where, EF_{Base} represents the EF of conventional asphalt pavement; $\text{EF}_{\text{Recycle}}$ represents the EF of recycled asphalt pavement; C_{Base} represents the cost of the conventional pavement; C_{Recycle} represents the cost of recycled pavement.

Criteria for Pavement Recycling

The SRI values help to develop criteria for pavement recycling. The three criteria for pavement recycling as follow:

1. Sustainable recycling $0 < \text{SRI} < 1$;
2. Limit of recycling $\text{SRI} = 0$;
3. Unsustainable recycling $\text{SRI} < 0$

The SRI value help to understand the limitations of recycling. For pavement recycling, the SRI value lies between 0 and 1, the pavement recycling would be sustainable in nature. However, for negative SRI value, pavement recycling would become unsustainable. For zero SRI value, it indicates to limit the pavement recycling. For improving sustainability in pavement recycling, the SRI value should be close to unity.

2.3 Economic Assessment

The economic assessment of asphalt pavement generally considers four parameters:

1. Cost of the materials,
2. Cost of the labour/manpower,
3. Cost of the machinery work,
4. Transportation cost.

In this study, transportation cost of materials, labourer and machinery has not been considered due to the large variation in data. The details of rest of the parameters are mentioned in Table 6.

Table 6 Details of parameters for economic assessment

Materials	Cost (Rs.)
Bitumen	26,000/tonne
Coarse aggregate	260/tonne
Fine aggregate	260/tonne
Waste plastic	12,000/tonne
Labour cost	500/day
Machinery	
Asphalt paver	1200/h
Rollers	619/h
Tack truck	42/km
Pickup trucks	1500/day
Small loader	968/h
Small broom	500/h

3 Case Samples

Amount of Ingredient for different types of sample in Marshal Stability Test Mould of 1030 g weight is as follows:

Convention Asphalt Pavement: For the conventional type of sample for the Marshal Stability Test Mould Assembly of surface course amount of course aggregate is used about 200 g, fine aggregate is used about 284 g, Dust is used about 492 g and bitumen is used about 54 g. And if we take the example of the one-meter cube the material required is coarse aggregate used is 339.5 kg, fine aggregate is 482.17 kg, Dust is 835.31 kg and bitumen is 91.68 kg. Samples images are shown in Fig. 5.

Asphalt pavement with 10% plastic: If 10% of bitumen is replaced by the waste polythene and it is cast in Marshal Stability Test Mould Assembly of surface course amount of course aggregate is used about 200 g, fine aggregate is used about 284 g, Dust is used about 492 g, bitumen is used about 49 g and waste polythene is used about 5 g. And if we take the example of the one-meter cube the material required is coarse aggregate used is used about 339.5 kg, fine aggregate is used about 482.17 kg, Dust is used about 835.31 kg, bitumen is used about 83.19 kg and waste polythene is used about 8.48 kg. Samples images are shown in Fig. 6.

Asphalt pavement with 20% plastic: If 20% of bitumen is replaced by the waste polythene and it is cast in Marshal Stability Test mould Assembly of surface course amount of course aggregate is used about 200 g, fine aggregate is used about 284 g, Dust is used about 492 g, bitumen is used about 43 g and waste polythene is used about 11 g. And if we take the example of the one-meter cube the material required is coarse aggregate is used about 339.5 kg, fine aggregate is used about 482.17 kg, Dust is used about 835.31 kg, bitumen is used about 73.00 kg and waste polythene is used about 18.67 kg. Samples images are shown in Fig. 7.

(a)   (b)

Fig. 5 Blocks of conventional asphalt for **a** surface course, **b** base course

(a)   (b)

Fig. 6 Blocks of asphalt with 10% replacement of bitumen with waste polythene for **a** surface coarse, **b** base coarse

Asphalt pavement with 30% plastic: If 30% of bitumen is replaced by the waste polythene and it is cast in Marshal Stability Test Mould Assembly of surface course amount of course aggregate is used about 200 g, fine aggregate is used about 284 g, Dust is used about 492 g, bitumen is used about 38 g and waste polythene is used about 16 g. And if we take the example of the one-meter cube the material required is coarse aggregate is used about 339.5 kg, fine aggregate is used about 482.17 kg, Dust is used about 835.31 kg, bitumen is used about 64.51 kg and waste polythene is used about 27.16 kg. Samples image are shown in Fig. 8.

Fig. 7 Blocks of asphalt with 20% replacement of bitumen with waste polythene for **a** surface course, **b** base course

Fig. 8 Blocks of asphalt with 30% replacement of bitumen with waste polythene for **a** surface course, **b** base course

Asphalt pavement with 40% plastic: If 40% of bitumen is replaced by the waste polythene and it is cast in Marshal Stability Test mould Assembly of surface course amount of course aggregate is used about 200 g, fine aggregate is used about 284 g, Dust is used about 492 g, bitumen is used about 32 g and waste polythene is used about 22 g. And if we take the example of the one-meter cube the material required is coarse aggregate is used about 339.5 kg, fine aggregate is used about 482.17 kg, Dust is used about 835.31 kg, bitumen is used about 54.32 kg and waste polythene is used about 37.35 kg. Samples images are shown in Fig. 9.

(a)   (b)

Fig. 9 Blocks of asphalt with 40% replacement of bitumen with waste polythene for **a** surface course and **b** base course

4   Results

4.1   Ecological Assessment

The Ecological Footprint of 1 km asphalt pavement (10 m width and 200 mm thickness of surfacing and binder coarse) is about 61.78 gha, however, the addition of waste plastic (i.e. for 10% waste plastic is about 59.30 gha, for 20% of waste plastic is about 56.82 gha, for 30% of waste plastic is about 54.34 gha and for 40% of waste plastic is about 51.86 gha) may decrease the environmental impact up to 16% of the total impact of the pavement. The details of the materials, labour and machinery use, etc. are mentioned in Table 7. The EF distribution of different asphalt pavement is depicted in Figs. 10, 11, 12, 13 and 14.

4.2   Economic Assessment

The constructional cost of 1 km asphalt pavement (10 m width and 200 mm thickness of surfacing and binder coarse) is about Rs. 5.87 million, however the addition of waste plastic may decrease the pavement cost up to 18% of the conventional pavement cost. The details of the different types of asphalt pavement are mentioned in Table 8. The distribution of the pavement construction cost is depicted in Figs. 15, 16, 17, 18 and 19.

Table 7 The details of the materials, labour and machinery use for asphalt pavement

Parameter	Conventional (gha)	10% waste plastic (gha)	20% waste plastic (gha)	30% waste plastic (gha)	40% waste plastic (gha)
Bitumen	25.40	22.86	20.32	17.78	15.24
Aggregate	5.27	5.27	5.27	5.27	5.27
Waste plastic	–	0.06	0.12	0.18	0.24
Storage	1.99	1.99	1.99	1.99	1.99
Mixing	24.60	24.60	24.60	24.60	24.60
Pavement placement	0.91	0.91	0.91	0.91	0.91
Labour	3.61	3.61	3.61	3.61	3.61
Total	**61.78**	**59.30**	**56.82**	**54.34**	**51.86**

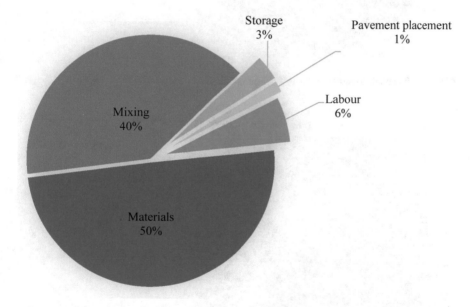

Fig. 10 EF of conventional asphalt pavement

4.3 Sustainable Recycling Index (SRI)

The SRI value of Asphalt pavement with 10% waste plastic use is estimated as 0.002, it increases gradually with the increase of waste plastic percentage in the asphalt pavement. The SRI of all proposed asphalt pavement is depicted in Fig. 20. The SRI value of asphalt pavement with 40% waste plastic is about 15 folds of the SRI value of asphalt pavement with 10% waste plastic.

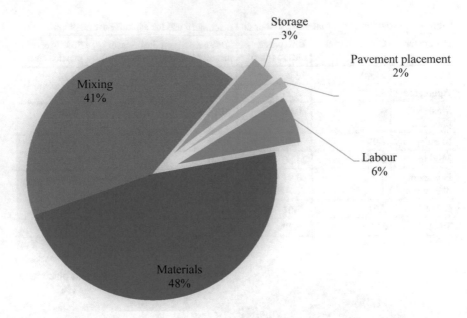

Fig. 11 EF asphalt pavement with 10% waste plastic

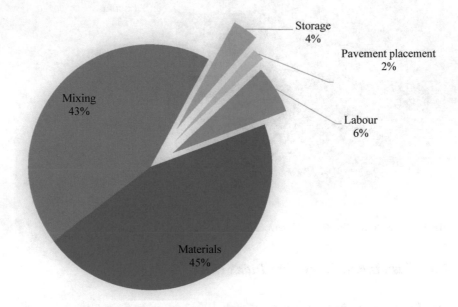

Fig. 12 EF asphalt pavement with 20% waste plastic

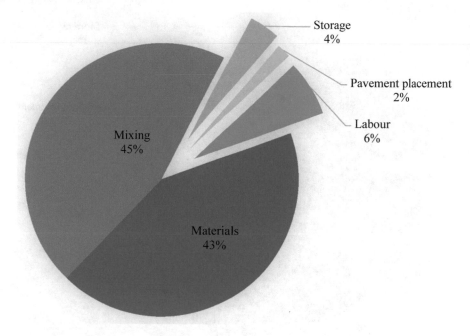

Fig. 13 EF asphalt pavement with 30% waste plastic

5 Conclusions

The Ecological Footprint of normal and waste plastic added asphalt pavement has been estimated and the results are compared with each other on the basis of pavement parameters. The EF of normal asphalt pavement is about 61.78 gha/km (with 200 mm surface and binder coarse thickness). The addition of waste plastic in pavement construction decreases the EF of the pavement because high energy density (bitumen) material is replaced by low energy density (waste plastic) material. Replacing 40% of the bitumen in asphalt pavement with waste plastic may decrease the EF of pavement construction up to 51.86 gha/km (i.e. 16% less than the EF of normal pavement). The EF of materials is maximum for the normal asphalt pavement, followed by pavement with 10% waste plastic and pavement with 20% waste plastic. The EF of hot mixing process become highest contributor for waste plastic addition of 30% or more in pavement. The results suggested that the use of low impact materials in the pavement decrease the overall EF of the pavement.

Use of waste plastic in pavement may also decrease the cost of pavement construction because the waste plastic cost (Rs. 12,000/tonne) is lower than the cost of bitumen (Rs. 26,000/tonne). The cost of normal asphalt pavement is estimated as Rs. 5.872 million per km pavement that is more than the cost of pavement with 40% waste plastic (i.e. Rs. 4.813 million/km pavement or less than Rs. 1.06 million). The cost

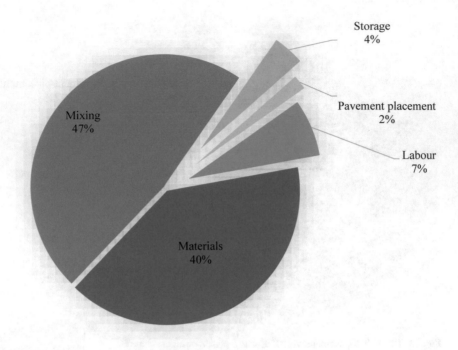

Fig. 14 EF asphalt pavement with 40% waste plastic

Table 8 The details of the cost of the different types of asphalt pavement

Parameter	Conventional (Rs. million)	10% waste plastic (Rs. million)	20% waste plastic (Rs. million)	30% waste plastic (Rs. million)	40% waste plastic (Rs. million)
Bitumen	4.91	4.42	3.93	3.44	2.95
Aggregate	0.86	0.86	0.86	0.86	0.86
Waste plastic	–	0.226	0.453	0.680	0.907
Labour	0.016	0.016	0.016	0.016	0.016
Machinery	0.081	0.081	0.081	0.081	0.081
Total	**5.872**	**5.607**	**5.343**	**5.078**	**4.813**

of pavement materials contributes to maximum pavement construction cost followed by machinery cost.

The SRI of pavement has also been calculated for the different percentages of waste plastic in normal asphalt pavement. The SRI evaluation demonstrates that the pavement with 40% waste plastic addition is the best of all the proposed recycled pavement. The addition of waste plastic in asphalt pavement is ecofriendly as well as economical also. In other words, the use of waste plastic helps to reduce the

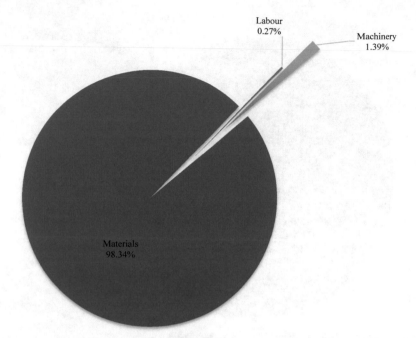

Fig. 15   Cost of conventional asphalt pavement

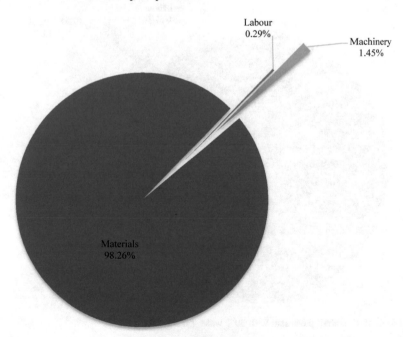

Fig. 16   Cost of asphalt pavement with 10% waste plastic

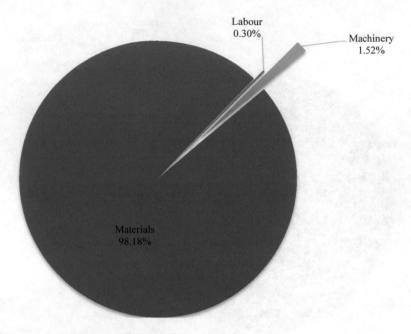

Fig. 17 Cost of asphalt pavement with 20% waste plastic

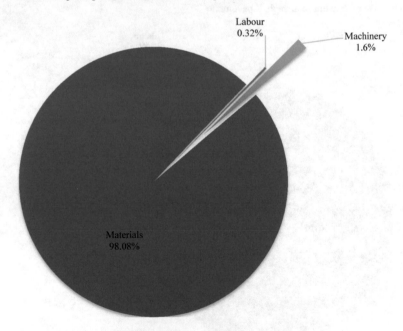

Fig. 18 Cost of asphalt pavement with 30% waste plastic

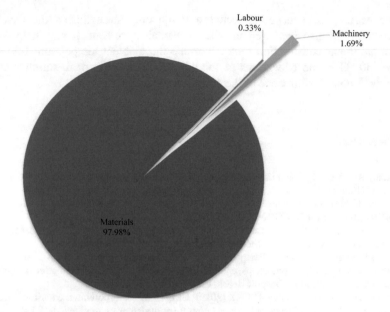

Fig. 19 Cost of asphalt pavement with 40% waste plastic

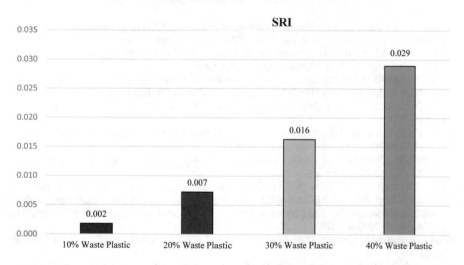

Fig. 20 Sustainable recycling index of different pavements

demand for bitumen, thus reducing the cost of pavement construction also. Recycled materials (waste plastic) can efficiently achieve the goal of sustainability if it used efficiently in pavement construction applications. It reduces the EF of the pavement construction and helps to prevent global warming related to high demands of asphalt for pavement.

The waste plastic recycling technology not only strengthened the pavement construction but also increased the life of asphalt pavement. It may help for the Indian pavement construction sector because the ambient temperature frequently crosses 50 °C in most locations of the country during the peak summer season resulting in poor performance of the conventional asphalt pavement.

References

Al-Hadidy AI, Yi-qiu T (2009) Effect of polyethylene on life of flexible pavements. J Constr Build Mater 23(1456):1464

Austoroads (2009) Guide to pavement technology: part 4E: recycled materials. Publication No. AGPT04E/09, Project No. TP1565, Austroads, Melbourne, Australia

Chambers N, Simmons C, Wackernagel M (2004) Sharing nature's interest: ecological footprints as an indicator of sustainability. Sterling Earthscan, London, Great Britain

Chehovits J, Galehouse L (2010) Energy usage and greenhouse gas emissions of pavement preservation processes for asphalt concrete pavements. In: Proceedings on the 1st international conference of pavement preservation, Newport Beach, California, pp 27–42

European Environment Agency (EEA) (2013) EMEP/EEA air pollutant emission inventory guidebook—2013. http://www.eea.europa.eu/publications/emep-eea-guidebook-2013

Federal Highway Administration (FHWA), U.S. Department of Transportation. Asphalt pavement recycling with reclaimed asphalt pavement. https://www.fhwa.dot.gov/pavement/recycling/rap/index.cfm. Accessed Jan 2021

Husain D, Prakash R (2019a) Ecological footprint reduction of building envelope in a tropical climate. J Inst Eng (India) Ser A 100:41–48. http://doi.org/10.1007/s40030-018-0333-4

Husain D, Prakash R (2019b) Ecological footprint reduction of built envelope in India. J Build Eng 21:278–286. https://doi.org/10.1016/j.jobe.2018.10.018

Husain D, Prakash R (2019c) Life cycle ecological footprint assessment of an academic building. J Inst Eng (India) Ser A 100(1):97–110. http://doi.org/10.1007/s40030-018-0334-3

Indian Horticulture Database (IHD) (2014) Ministry of Agriculture, Government of India. www.nhb.gov.in

Indian Roads Congress (IRC) "Guidelines for the design of flexible pavements", fourth revision (2018). http://www.irc.nic.in/. Accessed Jan 2021

Mannering FL, Washburn SS (2013) Principles of highway engineering and traffic analysis, 5th edn. Wiley, Hoboken, New Jersey

National Sample Survey Office (NSSO) (2014) Household consumption of various goods and services in India 2011–12. Ministry of Statistics and Programme Implementation, Government of India. http://mospi.nic.in/sites/default/files/publication_reports/Report_no558_rou68_30june14.pdf. Accessed 22 Nov 2020

Open Government Data (OGD) (2019) Government of India. https://data.gov.in/node/94765/download

Reid M (1998) ALT-MAT: alternative materials in road construction. In: Polluted+ marginal land-98. Proceedings of the 5th international conference on re-use of contaminated land and landfills, 7–9 July 1998. Brunel University, London

Scripps Institution of Oceanography (SIO) (2017) The keeling curve. https://scripps.ucsd.edu/programs/keelingcurve/2013/07/03/how-much-co2-can-the-oceans-take-up/. Accessed 23 Mar 2020

Vansudevan R, Sekar ARC, Sundarakannan B, Velkennedy R (2012) A technique to dispose waste plastics in an ecofriendly way—application in construction of flexible pavement. Constr Build Mater 28:311–320

Wackernagel M, Rees W (1996) Our Ecological footprint: reducing human impact on the earth. New Society, Gabriola Island, British Columbia

Printed in the United States
by Baker & Taylor Publisher Services